FIX IT BEFORE IT BREAKS

FIX IT BEFORE IT BREAKS

Seasonal Checklist Guide to Home Maintenance

Terry Kennedy

McGraw-Hill

New York Chicago San Francisco Lisbon London Madrid
Mexico City Milan New Delhi San Juan Seoul
Singapore Sydney Toronto

The McGraw-Hill Companies

Cataloging-in-Publication Data is on file with the Library of Congress.

1 2 3 4 5 6 7 8 9 0 DOC/DOC 0 1 0 9 8 7 6 5 4

ISBN 0-07-143069-5

The sponsoring editor for this book was Larry S. Hager, the editing supervisor was David E. Fogarty, and the production supervisor was Pamela A. Pelton. It was set in Melior by Joanne Morbit and Paul Scozzari of McGraw-Hill Professional's Hightstown, N.J, composition unit.

Printed and bound by RR Donnelley.

This book was printed on recycled, acid-free paper containing a minimum of 50% recycled, de-inked fiber.

This Book Is for My Dad

CONTENTS

PREFACE

Rot, boring insects, molds, fungus, and earth movement are some among the long list of forces that attack your structures over and over. Repair of the damages caused by the onslaught of nature costs building owners hundreds of millions of dollars every year.

Even if you haven't paid for repairs yet, the bills will catch up with you, or your estate. When you are retired and want to sell and travel on the equity in your buildings, or put the funds into a business, or pass title on to your children, termite work and major structural repair can rear their heads and take a huge bite out of the financial security, that you thought you had secured through all those years of mortgage payments.

And to make matters worse, if there are lawsuits against the real estate for construction defects or errors and omission, they are long, drawn-out and very expensive—whether you are being sued or have to file an action against parties whom you alleged wronged you.

The repairs and the litigation cost the United States billions of dollars each year, and the costs just continue to escalate. There is nothing that can be done about the forces of nature—and we wouldn't want to lose the natural abundance and beauty—we have to have rain and wind and earth. But taking care of your home in a diligent manner can shield you from a lot of expenses. The straightforward steps in this book can save a building owner huge amounts of money simply by prevention of damage to the property.

This is not a handy person's book. Hundreds of books can coach you on how to fix anything around your buildings. We list websites in the Resource Directory at the end of each chapter, and from the World Wide Web you can search for books on any repair subject.

The goals of *Fix It Before It Breaks* are very simple:

- Set up a seasonal checklist system that you understand and that you can use to keep your home maintained.
- Help put you in a position so you can know whether the correct maintenance is being done and what questions to ask if you don't understand what's happening.

With *Fix It Before It Breaks*, you can develop a simple maintenance program to save your family hundreds of thousands of dollars over the next few generations—and you can do it by yourself, with the help of your chosen group of experts; or you can turn the whole thing over to your architect or builder.

Our goal is to make this work simple and fun rather than heavy and tiresome. After all, it is always more enjoyable to know what you are doing than to be led along by a leash of self-doubt and confusion. We wish you the best and feel certain that you will gain a lot of satisfaction from knowing that you are actually taking care of your home in a very important way.

Visit our Web site today: http://www.safehomesweethome.com

Terry Kennedy

ACKNOWLEDGMENTS

I want to offer my thanks to some people who have been there for me on the day-to-day level, where it really counts:

My mom and dad, the best.

Ms. Lynn Garlick, a great writer of short stories and the kind of true, honorable, tireless friend who makes life worthwhile. Thanks, Lynn-you are a never ending boost of energy with the best feedback a person could possibly hope to find.

And—a team of *barristi* from Long Beach, California—the staff who hosted at Starbucks café #5255. They transform the United States by hanging in there day-after-day and providing us that fine sense of community, the café ambience, that one finds in southern Europe.

William, Bill, Kristiana, Patrick, Steve, Cynthia Ann, Michael, Vanessa, Kristin, Michael, Idaly, Shrah, Michelle, Teressa, Toni, Sarah, Lindsay, Emilio, Joanne, Thu Truc, Jessica, and Julie.

ABOUT THE AUTHOR

Terry Kennedy is a construction analyst in massive construction lawsuits with more than three decades' experience as a residential builder and remodeler in all parts of building: earthwork, concrete, and framing, with all types of homes.

CHAPTER 1

Getting Started

Building owners lose hundreds of millions of dollars to unnecessary repairs, increased insurance costs, and lawsuits every year. Right now, as you read this, everyone's home is under attack from common, natural forces, and most people have no idea how vulnerable they are. Take a look at Fig. 1.1. At a quick first glance, it appears that a swipe of caulk and some paint will fix things right up.

People stroll right past serious problems that would never have existed if a simple maintenance routine had been in place (Fig. 1.2). Every moment, day and night, natural forces continue without a break. Hidden destruction can be found in many places, and around the clock the onslaught from nature never lets up. Naturally, we would not want her to go away because she provides so much for us, but what is the ordinary person to do about this serious situation?

Who This Book Is For

Fix It Before It Breaks is not just for handy people who know how to make home repairs—*it is for everyone.* The busiest, least skilled people in the world can use the checklist system to understand what needs to be done, what professional maintenance people are supposed to be doing, and to keep their property safe.

POINTERS

Just as with an automobile, an envelope full of maintenance checkoffs and repair receipts can be an excellent tool whenever one is interested in selling a real estate parcel.

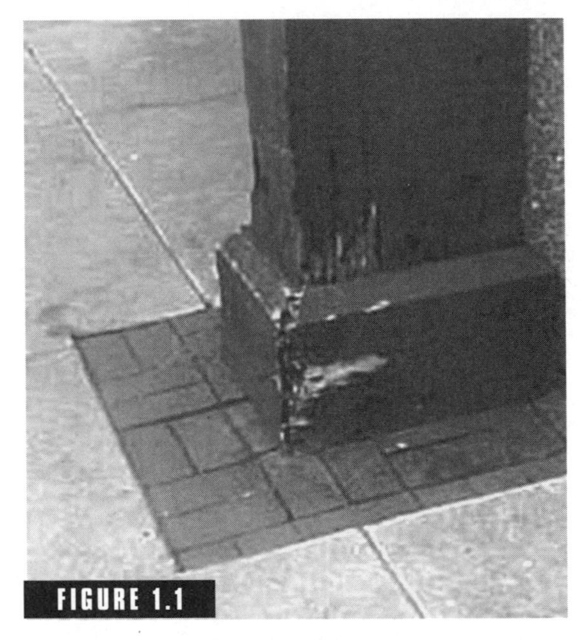

FIGURE 1.1

Hidden damage costs owners a tremendous amount of money. You may notice damage, but it doesn't look too bad so you walk by. To view what's below the surface, see Fig. 1.2.

- *Busy building owners.* You care about your real estate investment and are strapped for time. You can hand the book over to an architect or builder you trust, who can take it from there. But now you will know how to monitor their work.
- *Estate planners.* Get a grip on the deterioration of real estate holdings. Damages from poor upkeep are a very serious, ongoing burden—repair bills can be immense.
- *Homeowners with no knowledge.* Skim-read the book, have your architect or builder work up the checklists, and engage a handyperson, or a maintenance company to follow through on your home's individual needs. You will have the advantage of knowing whether they are doing things the right way and when to call other parties for advice.
- *Owners of multiple properties.* Your architect can monitor the physical work through the years, so you will be extra safe and will not have to pay attention to the program at all. But you will know how to ask pertinent questions about the welfare of your real estate.
- *Hands-on types, handy people, or people who just love their homes and find fixing things a welcome change of pace.* You can use this book and learn what really has to be done.

Regardless of how you address the needs of your buildings, by using this checklist system, you will finally be in a position to know for sure what you need to watch out for, whether professionals are informed, and if the maintenance people are actually doing a good job.

You will also learn when to call an expert if you are in doubt and how to incorporate the expert's feedback into your own, personal home care checklist system.

Understanding Your Home's Needs

First, you have to be aware of what is really happening. After the trim boards are removed in Fig. 1.2, the damage and the large, looming repair bills become obvious. Because the care of homes is so crucial, you must learn to put reliable protection programs together and follow them with diligence, just as you do with your cars.

We will show you Fig. 1.2 again, in Chap. 10, where you can get a sense of just how much you have learned about the hidden dangers that lurk in your home.

Remember, the goal here is not to become stressed out and panic—home maintenance can be an enjoyable, fulfilling experience, and we will help you get some pleasure from the process. Fear and worry are not necessary. What is important is to learn how to provide the TLC our homes require.

With cars, we get a manual, and then we take the vehicle to our mechanic to establish an upkeep plan. But with buildings, which for most people are far and away the most expensive purchase of a lifetime, there are no safeguard systems in place. Architects, builders, developers, and maintenance companies—none provides you with a manual or any other kind of guidance. To make matters worse, there is no way of knowing if a maintenance company is capable of providing professional service, and it is astonishing how many lawsuits are actually triggered by poor upkeep.

Call the Doctor

Never wait. If your intuition tells you that something is wrong, no matter what—water stains, earth movement, or a sparking arc at an electric terminal—get out your Who Does What Checklist. It's the first one you will make (later in this chapter). If you have doubts about who to call, for example, your concrete expert or your soils engineer, don't wait. Just call your general contractor or architect right away.

FIGURE 1.2

When severe destruction is uncovered, it is a shock, but diligent maintenance can prevent damage. Go back to page one and take a look at Fig. 1.1. The roof will have to be shored while the posts are replaced. If the trim boards at the base of the post had been caulked consistently, the paint preserved, and regular pest inspections taken place, these repairs could have been prevented.

This book is designed to fill the void in your maintenance knowledge simply and quickly. We do not discuss simple items such as paint, switches, light fixtures, or appliances that simply break down. We address only areas of a home that can have a serious effect on other parts. For example, we don't cover defective dishwashers, but we do advise that you watch for plumbing leaks. This

POINTERS

A tidal wave of lawsuits related to homes is sweeping the United States: errors and omissions, homeowners, construction defects. They take up a lot of a family's time, and many do not even pay for fixing the damages. But even if a settlement does not pay for everything, a diligent maintenance system can cut the repair costs so significantly that you are not devastated by an unfair outcome.

Below is a typical item from a discovery list of defects in a subdivision. This cost of repair is just for one door on the homes, and there are 200 homes in the subdivision. This will give you some sense of how much all the repairs on your home can add up to for doors, windows, foundations, etc.

If you do the math, you see right away that fixing an entire subdivision can require millions of dollars. Even if the owners didn't get a fair payoff, simple maintenance would have kept their homes in better shape.

5.6 Water entry at side doors

Defect: **The side doors to concrete porches and verandas.**
Frequency: **14 homes inspected, 14 incidents = 100% defective**
Damage: **Major mildew and mold infestation; damaged framing lumber; damaged framing metal; decayed floor sheathing, pad, tack strip, and carpet; door frame components damaged, door coming unglued at styles, all paint and finishes destroyed: all are the result of defective work product at doors.**
Suggested repairs: **Remove all carpet, door jambs, thresholds, and needed sheetrock and finishes. Replace all damaged material with correct sealants and polymer building paper. Provide construction details and specifications by architect in charge.**

Cost of Item: $4780.00 per home

is not a handyperson's or a home inspection book—it is a prevention system book, instructions for developing a maintenance plan contingent on checklists, which can eliminate the need for many very expensive repairs. It is structured to put you at the helm, so you are fully capable of taking charge of the protection of your home smoothly, simply, and continually.

Your Goals Are Easy and Straightforward

1. Starting with the land it sits on (Chap. 2, "The Building Site"), you will do the Quick Scan in each chapter. This is a handy inspection

that helps you learn more about each part of your building.

2. After the Quick Scan, you will fill out the appropriate checklists for each part of your home as you proceed through the chapters. These surveys help speed up the learning curve and make it easier to see your structure as a whole unit.
3. In Chap. 10, "Developing Your Personal Maintenance Plan," you will use all the knowledge from your previous lists to complete the final, seasonal checklists. They flag exactly what needs to be done every year.
4. Use professionals. The book helps guide you about when to get help and how to choose the right person to aid you with any checklist that is confusing.

POINTERS

Have you ever stood there with contractors proposing work on your home and wondered if they knew what they were talking about? This book shows you when to call for expert opinions and how to grasp what the plan of action means. Understanding the details is what puts you at the helm—and these "Pointers" sections appear throughout the text.

If you do not have a precise system in place, it is highly probable that inadequate care of buildings will encumber your family with hundreds of thousands of dollars of debt within the next couple of generations. Even if it is your children or their youngsters who sell, the estate will be required to fix any defects or be yoked with massive litigation.

The data are delivered in a structure that is designed to inform you about needed procedure quickly. If you apply a little effort and some observation, you can learn to take care of the hurdle of the ongoing deterioration for good. Even if you prefer to hire someone to do it all for you, this book can prepare you for monitoring and questioning the professionals you choose.

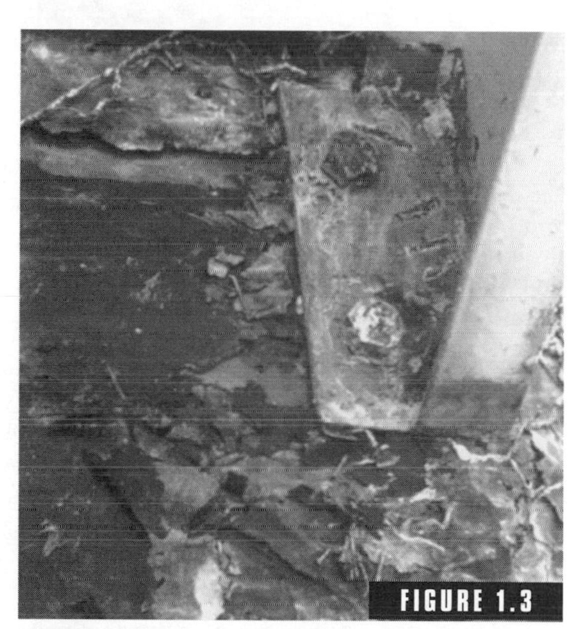

FIGURE 1.3

First the building paper rots and rust begins—lack of maintenance often leads to serious unseen repairs. Next, the structural steel rusts through and framing lumber rots. Even if you live in a condo or a subdivision with contracted maintenance, you must keep an eye on the professionalism, diligence, and quality of the company's work.

You have undoubtedly walked past areas of your home similar to the stucco crack in Fig. 1.4 and felt anxious, then carried the problem around in your head. You knew that something might be wrong, but did not know what. You wondered if something should be done, but you didn't have the time to investigate and did not know whom to call for a quick understanding of what should take place. *Fix It Before It Breaks* will give you control over these types of situations.

FIGURE 1.4

Serious defects are often neglected. Homeowners see problems, but overlook them because they don't know how much damage they cause. The photo depicts the upper right corner of a window. The aluminum frame of the window is separated, and there is a crack at the siding, above the window frame. Both will allow water to leak into the structure.

After designing your own personal plan with professionals, you can use any reliable individual to follow through with the work required to protect your properties. Or, you can take care of the simple jobs yourself, hiring others to carry through with the more complex jobs. The majority of regular chores are not very difficult, and you might actually find it soothing to roll up your sleeves and put a little elbow grease into your home. The beauty of *Fix It Before It Breaks* is that *you* are in charge, no matter how you plan to take care of your buildings.

What You Will Find in the Book

Book Layout

We start with the site—your yard and the rest of the lot on which your buildings are built. Then we move on to explore your structures from the bottom at the foundations to the top of your roof, just as it was built. This chapter, "Getting Started," walks you through the big picture of how to use the book and how to think about what you are doing when you put together a home protection plan.

Beginning with Chap. 2 and all the way through the book, we provide readers who prefer not to wade through a lot of words with a jump-start section. Right after a brief opening note that introduces the chapter, the "Jump-Starting" section fills you in on the important parts of upkeep for this portion of the project. The information is designed specifically for those who don't want to read the details but just want enough to get going and then (perhaps?) come back and read more as time goes by.

After "Jump-Starting," we move into developing the checklist. Photographs and drawings are interspersed within all the sections to provide clarity. This sets it up so anyone can jump-start and then develop checklists unique to a building.

We focus on the lists for the heading and then, at the end, provide the "Bringing It All Back Home" section, which ties up jump-starting and developing the checklist. Then, the reader is ready to move on to the next chapter.

After the "Bringing It All Back Home" section, "The Resource Directory" opens the information spigot wider still. Since this book is not designed to be a how-to-fix-it manual, but is specifically oriented for the building owner to develop a maintenance checklist system, "The Resource Directory" helps you to find more information. Some owners may never get their hands dirty. But if you choose to do the work, there are dozens of how-to books in the marketplace.

A simple trip to the bookstore or a quick search online will offer up many home repair titles for doing your own work, and "The Resource Directory" guides you toward still more information about subjects associated with the checklists.

In an effort to make "the Resource Directories" appropriate twenty-first century databases, we guide you to sources on the Web that will help to clarify and lead you to further repositories. The HTMLs are in place so you can simply type them into your browser line and go right to the source.

The Headings Look Like This

Jump-Starting

The goals for the chapter are listed right in the beginning of each "Jump-Starting" section. Flip through to any area of the book, and you will find this heading. For example, in Chap. 2, "The Building Site," there are two main objectives:

1. Learn to make sure that water is not causing problems.
2. Understand how to determine if the earth is stable.

THE PLAN

After setting the goal, you leap right into the plan and can begin to get an exact idea of what you are supposed to achieve in the chapter. Here, again we use Chap. 2 as an example.

1. Check out the site for stability.
   a. If the earth is not stabilized, fix it.
   b. When it is stabilized, keep it that way with diligent maintenance.
2. Check the site for drainage.
   a. If water is not running off the site, drain it immediately.
   b. Keep it draining with regular maintenance.
   c. Check for signs of underground water.
   d. Repair detrimental groundwater problems.
   e. Monitor groundwater repairs.

The Quick Scan

These sections are for taking a rapid tour of the part of the building being scrutinized. In some cases there are various types of systems used for different buildings. For example, some buildings have T foundations, while others have a simple footing with a concrete slab. "Quick Scan" takes you through the considerations that you must explore in order to develop your own personal maintenance plan. The "Jump-Starting" section, in which "Quick Scan" is located, is designed to achieve two things:

1. Tell you what to look for at that particular part of your building and what to watch for in that area.
2. Guide you through the process of turning your new knowledge into the final, ongoing checklists that begin with the individual parts of the parcel of real estate, then bring it all back together for an overall list that becomes your ongoing plan.

Open to this section in any chapter, and you will find information on how to get going with the particular phase of maintenance that you are ready to work on. Look at the steps, grab your pad and clipboard, and do exactly what the "Jump-Starting" section says—the text is self-explanatory and straightforward.

Jump-Starting Your Use of the Book

Think of your buildings as a complex of systems: the earth; the grounds; the foundation; the skeleton; the skin with doors and windows; the roof; the plumbing, electric, and HVAC (heating, ventilating, and air conditioning) systems; and the interior.

Keep this simple view of your home in mind. In the "Jump-starting" sections, we are going to go through a part of your building, guiding you to each area that is in need of attention and highlighting the situation with photographs and line drawings.

Try it out with Chap. 2, "The Building Site." The steps for jump-starting are simple:

1. Skim-read the "Jump-Starting" and the checklists sections in Chap. 2.
2. Follow the directions for doing your own survey of your site. Jump-Starting is a rapid approach to a method, but this is critical work and you must take enough time to understand what you are

doing. If it looks too hard, or as though you do not have the time, retain an architect or a builder who will understand the material quickly and can make it happen fast.

3. Develop your building site checklist and review it with the jump-starting information.
4. Go back to the information again if you are confused.
5. Bring in a professional if you have questions.
6. Edit and rewrite your checklists.
7. Move on to Chap. 3, "Foundations and Cement Work."
8. Follow all the steps above.
9. Continue with Chap. 4, "Framing," and move on through all the parts of your structure and the grounds.
10. In the final Chap., 10, "Developing Your Personal Maintenance Plan," you will pull all the information together.

POINTERS

Does this sound as if it's all a bit much? Your schedule is just too hectic. You can't possibly put more tasks on your calendar—they just won't fit. *Remember,* this system is set up so others can do all or parts of it, *and* you will know how to ask questions when you wonder if they are doing a good job.

Fill out the checklists and you will have developed a safeguard system unique to your structures. There is a monthly list, and the routine for each season will be there waiting for you at that particular time of the year.

Developing Checklists

Remember that the various lists in the chapters about specific parts of your building, for example, roofs, doors and windows, are laid out carefully to achieve three things:

1. Help you locate the areas around your property that require care and get a grip on where you need to provide diligent maintenance to protect your home from unnecessary wear and tear.
2. Guide you in determining when to call in an expert, how to use that person to determine what needs immediate repair, getting it fixed, and maintaining the part of the building after the repairs are completed.

POINTERS

To simplify this project, go through all the chapters first. Note anything that you are not sure of, and list your questions. *Then* get an expert for the things you don't understand. Determine what needs fixes or repairs. Have the work done, and ask the repair people for maintenance tips—most will be very helpful. Then proceed with finalizing your maintenance plan, using the ongoing maintenance lists in Chap. 10.

3. How to systematically use the series of checklists you build in the early chapters, bringing the information back together to create the checklists that you will use for the final, ongoing maintenance plan. Through the years you will establish a crucial, home protection program that can be handed down to future generations.

Below is your beginning checklist. At first glance, the subject matter may sound trivial, but rest assured, these contacts are very important. Gather the information as soon as possible—it is never a good idea to be in need of a trustworthy professional when you are in a pinch. Always try to avoid choosing a person for any job at the last moment.

There is no better way to find the people you need than by word of mouth. Whether they be societies, organizations, bureaucracies, or for-profit listings, no other source is like the firsthand experience of your friends and neighbors who have dealt with contractors and workers in their own homes. Not even the endorsement of friends can guarantee that you will have success with the individuals you hear about, but your chances will be greatly improved.

Start filling out this list right now. Work on it as you read the book. Ask friends and family for contacts if you do not have someone lined up for a trade and service already. Note that we left space for you to add services that may be unique to your home.

Who Does What Checklist

Trade	Notes
Architect	
Attorney (for construction)	
Attorney (for real estate)	
Building department	

Trade	Notes
Carpenter	
Concrete contractor	
Drain opening company	
Electrician	
Estimating company	
Electric vendor	
Engineer—soils	
Engineer—structural	
Fire department	
Gardener	
Gas company	
General contractor	
Grading (earthwork) contractor	
Handyperson	
Hardware store	
Heavy equipment rental	
HVAC (heating, ventilating, and air conditioning) contractor	
Laborer	
Landscaper	
Lumberyard	

Trade	Notes
Maintenance company	
Maintenance person	
Painter	
Plumber	
Roofer	
Septic system	
Stone, brick, and block mason	
Tool rental yard	
Tile setter	
Trenching contractor	
Underground detection	
Waterproofing contractor	
Window supplier	
Others	

Bringing It All Back Home

This is the last section in each chapter. It includes a bit of extended reference material that you can come back to time and again, read slowly, and digest at your leisure. Where required, the descriptions go into detail about why you are performing the various activities. The photography and drawings help clarify what to do and what is beneath the surface, and they reveal examples of what happens if you are not persevering with care of your house. "Bringing It All Back Home" works toward two basic goals:

1. Add a bit of depth and breadth to the information.
2. Tie up the chapter in order to segue into the next subject.

Other Features

Pointer Boxes

Found throughout the book, "Pointers" are tips for the reader to use in fine-tuning their knowledge about a building. The details cover information that would be learned through the time-consuming process of trial and error, and they can save you time and help ensure that the work is done correctly the first time.

"Call the Doctor" Boxes

The "Call the Doctor" box will advise you to call an expert from your list. These boxes are placed where it is critical that you find out if ongoing, compounding damages are destroying the structure. For example, suppose an ongoing, moist earth condition is found at a

POINTERS

Throughout *Fix It Before It Breaks*, the "Pointers" sections guide you with specific advice. They will help you get a handle on your home maintenance system. The tips make diligent maintenance much simpler and less expensive than waiting until there are problems and having random repairs done.

Helpful and interesting information boxes are found in every chapter. This book is about building quick, easy checklists to safeguard your home from the elements—it is not a fix-it encyclopedia or a repair manual. Most any subject that you wish to explore in depth can be found online by using the *Fix It Before It Breaks* Resource Directories.

 **Call the Doctor** 

These boxes are definite warning flags—when in doubt, *bring in a pro.* Whenever we come to a serious situation anywhere in the book, anything that could harm your home or family, you will find a "Call the Doctor" sign post.

corner of a building and there is no clear method for determining the source of the moisture or its effect on the framing. Building corners can carry a lot of weight from the materials above, and if the foundation is supported by soggy soils, there is opportunity for the structure to settle, which can cause serious damage. And if the leak continues to go unattended, more of the foundation, hardscape, and flatwork can be undermined.

Suggestions about when to call an expert are not only found in the "Call the Doctor" boxes—we have included the thought that it would be wise to contact a person with deep knowledge in numerous places. If you are knowledgeable, please be patient; a great many of the people who use this system have little or no understanding of the subject. It is to their benefit to use skilled advisors wherever possible.

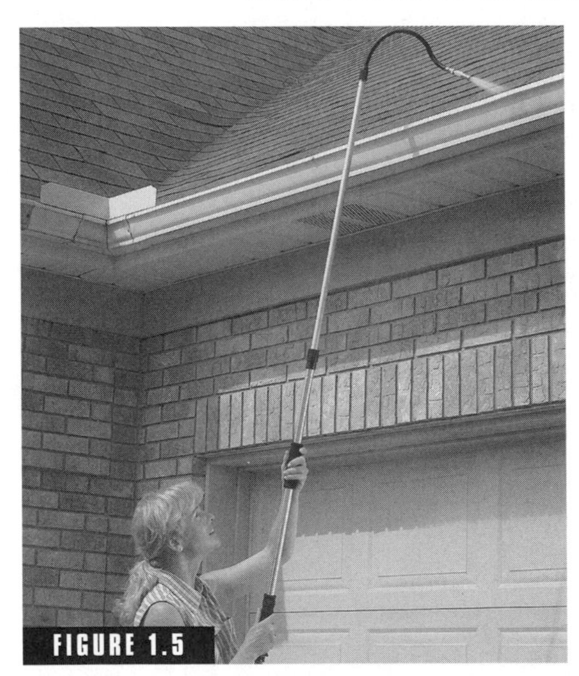

FIGURE 1.5

It's important to put "clean-out gutters" on your to-do list, because expensive repairs await if your gutter-and-downspout system doesn't drain properly. Attach your garden hose to the flexible "Water Snake" shown here to clear leaves and other debris from gutters while you stand safely on the ground! Visit the Improvements Web site (improvementscatalog.com) to see this and other clever home-maintenance items!

In General

The entire book is loaded with information that will help you enjoy the process, make the jobs easier, broaden your knowledge, and generally make you much more efficient at taking good care of your real estate. For example, Fig. 1.5 introduces you to the Water Snake, a nifty tool for the fall, when gutters require extra attention if they are filling up with leaves and water at the same time. The miscellaneous information, as in all areas where it is applicable, includes the Web site address so you can go to it easily.

Developing the Final Maintenance Program Checklists

The preliminary series of lists is the backbone of the entire project. You will go through all the chapters—starting with the important "Who Does What Checklist" above, in this chapter—and fill out your individual pages.

a quote on any construction material. If you need complex repair vork, this site may help in monitoring the estimate you receive. 'ou need to be a true hands-on person or enlist the knowledge of an xpert. Construction estimating is very complex and requires xtreme diligence.
ttp://www.get-a-quote.net/QuoteEngine/costbook.asp?WCI=Cost

lthy House Institute books and videos. Learn more about toxins in ur buildings.
ttp://www.hhinst.com

ne information. This site contains more excellent information.
ttp://www.designcoalition.org

netime.com. This is a good source of information.
ttp://www.hometime.com

ise information. More information about your home.
http://alsnetbiz.com

wstuffworks.com. Excellent information about a number of home subjects.
http://www.howstuffworks.com

w your house works. This site is just what it says.
http://www.hometips.com

provements Catalog. Improve Your Home Sale and Discounts Premier Shopping Affiliate Partner.
http://www.ImprovementsCatalog.com

line coupons for your home and garden needs. Visit Coupon Mountain for links to all the latest coupons for home furnishings and garden supplies.
http://www.couponmountain.com/data/category.php3?a ctCat=19&s

viceMagic.com. This is a list of contractors for all types of work.
http://www.servicemagic.com

nley Works. Here is an old standby in the light construction tool business. Catalog lists tools and hardware for home improvement, consumer, industrial, and professional use.
http://www.stanleyworks.com/

rk clothes—protect yourself. It is imperative that you wear this Australian headgear while studying your site. Skin cancer is on an alarming rise in the United States. The Aussies know what to do—get hats that really work. These quality hats provide maximum sun protection, and they are stylish. With the breakdown of the atmosphere, uv radiation is very serious. Do not take a chance

Next, in Chap. 2, you will do the Building Site Checklist. Fill it out and proceed all the way through, working from the earth up to the roof. As you proceed, you will begin to get familiar with each part of your building.

Overall, the amount of your time required is minimal—a few hours as you get started and then simple maintenance every year, unless you hire someone to take care of the work. It does not take a long time to get started, but it is very important that you not rush through as you get to know your home and your neighborhood. Allow yourself plenty of time to look at things very carefully. Have a group of friends join you and discuss your opinions of the information at each person's home.

POINTERS

Resultant damage is destruction to your home that is consequential to defective work product from real estate developers, contractors, and subcontractors. If the developer fails to monitor the work of others and the bad work wrecks part of your home, it is called resultant damage because it results from poor work.

The Quick Review in Chapter 10

In Chap. 10, "Developing Your Personal Maintenance Plan," you will be taken through a review. For your convenience, we present a simple outline of what you have learned. Then you can rewalk your home, call up a professional, review any of the checklists in detail, or simply run through the review and use the knowledge you have accumulated to fill out the six checklists that you will use on a yearly basis as part of your personal plan.

The Final Six Checklists

- Keep an Eye Out Checklist
- Monthly Checklist
- Spring Checklist
- End of Summer Checklist
- End of Autumn Checklist
- Midwinter Checklist

If you have ever been involved in a lawsuit related to the sale of a structure, locked into construction defect litigation, or had to spend a big bundle of your profits on rot, mold mitigation, and termite repair when you sold a home, you will understand what this book is about immediately. Prevention can save lots of money.

FIGURE 1.6

Resultant damage is very serious. There is another view of this timber in the beginning of the chapter. It is a structural member—notice the rusted bolts. When it fails, it could cause bodily injury. This is an excellent example of unseen parts of your building being destroyed from lack of maintenance.

The United States is facing a tremendous problem. An endemic assault on our built structures is overwhelming the courts and accruing billions of dollars' worth of preventable repairs yearly. See Fig. 1.6.

The final six checklists are there for you to bring all the checklists together and establish your ongoing strategy. With list coordination and diligence, the building owner can prevent deterioration and costly repairs. The guidelines for the rest of the book are brought together in a comprehensive, simple manner—the reader can get a handle on the site and buildings as interactive systems.

> **POINTERS**
>
> **You can use a search engine on the Internet to find contractors. You will need some patience to find these companies in your vicinity—try doing a search for search engines, then run your searches on several of them. Don't rush the process—some sites that list contractors include additional information related to many parts of construction.**

The Resource Directory

At the end of every chapter you will find "The Resource Directory". We start right here in this chapter—the general information section is listed below. This book is designed specifically to create a checklist system for taking good care of your property, not to study the quality of products and services. Therefore, we do not guarantee or promote any item or group in the list—they are simply provided for your interest and convenience. The best assurance for the quality of a product or provider of services is typically a recommendation by a trusted friend or professional. There is still no guarantee that your experience will be fruitful, but it can help to know that the person or product has performed well with someone you trust.

All the Resource Directories include the HTML addresses that are used on the World Wide Web, so you can go right online and use the Web site in a twenty-first-century fashion.

If you want to fix things yourself, you can also find how-to books on a Web-wide basis. Just use a search engine whenever you want to try a repair on your own and you will find tons of information. Use the

McGraw-Hill catalogue, online book
bookstore with a large home section
fix-it books.

The Getting Started Resourc

American Technical Publishers Book
want to do their own fixes and m
information.
http://www.go2atp.com

Askbuild.com. Goes into many aspects
as other construction information.
http://www.askbuild.com

Build.com. This construction directo
mation about products, retailers
providers.
http://www.build.com/

Click Here for Hand Tools. Snap-On Te
quality hand tools, power tools, too
automotive tools, and much mo
http://buy1.snapon.com

Construction Defects. A construction
look at problems that homeowners
http://www.constructiondefects.co

Construction information. This is a goo
http://www.bobvilla.com

Contractor Guide—Find a contractor.
760,000 construction contractors, 1
manufacturers.
http://www.ContractorGuide.com/d

DoItYourself.com. This is an Internet c
How to's, Q&A bulletin board foru
household hints, and more. When
Yourself.com" is their slogan.
http://www.doityourself.com

Farm structures—elements of constructi
full of very good information. It is a
but the technology might save you m
http://www.fao.org

on getting skin cancer while you are doing your site and building surveys and working up your lists.
http://www.schoolsunhats.com

Workwear at Duluth Trading Company. Innovative clothing designed for builders, contractors, electricians, framers, carpenters, gardeners, and anyone with the need for comfortable workwear and clothing.
http://www.duluthtrading.com/link.asp?shop=catclothing.asp&s

CHAPTER 2

The Building Site

The building site is the land on which your home is situated. It includes the yard and the surrounding terrain. The other parcels near your home and even the streets—they are a drain system—are also of significance to the condition of your own property.

Taking care of the land that supports your structures is very important. There are many lot-related conditions that can cause damage to dwellings on a year-round basis, for example, water draining toward foundations and undermining them, rusting metals, earth buildup against siding, leaking underground drains, and invading termites.

This is a critical chapter. It is long and requires close attention, but bear in mind that it is well worth the effort. Site repair can cost homeowners more in out-of-pocket expenses, and draw owners into more complex lawsuits, than any other part of the whole real estate parcel. So relax with this chapter and pay close attention. If you have trouble, go back and review it again.

If you still have questions and/or are suspicious of any water, dampness, or earth movement (see Fig. 2.2) anywhere in the vicinity of your home—even if it is off the property—call an expert immediately. Putting off an investigation can cost you a lot of money in the long run. While you wait for the meeting, move on to the "Jump-Starting" section.

This chapter can provide real help in understanding how your sites work. You will use the information you gather for the Building Site Checklist to prepare your overall maintenance plan in Chap. 10.

Sprinkler components can leak. Note the ponding water at the arrow. The black behind is waterproofing, but the water can soak the soil beneath the foundation and undermine it. Sprinklers flooding foundations and lower parts of the framing are common. Usually they just spray the siding and leak into the bottom of the frame, causing rot and mold. However, there can also be underground leaks that flood the earth, causing foundation failure.

Jump-Starting

Starting your site maintenance survey is straightforward. There are three main tasks involved with keeping the lot stable and dry—we have used them for the layout of the Building Site Checklists, which you will see later in this chapter:

- *Your neighborhood.* Tour the neighborhood for drainage and soil movement that affect your parcel of land.
- *Drainage.* Make sure that surplus water flows away from your land readily.
- *Soils stability.* Monitor and control the movement, buildup, and/or erosion of earth.

First, take a close look below at the plan for compiling your personal Building Site Checklist. Then, in the "Quick Scan" section, we will explore the neighborhood to understand how it drains. Next, you walk your lot, then start filling in your checklist. If you have any confusion as you compile the information, be sure to call in an expert.

FIGURE 2.2

Check closely for a 6-inch clearance at foundations. Notice how earth in the corner, to the right of the pile of bricks, has built up against the bottom of the siding at the concrete foundation (you can see concrete above the earth right behind the brick pile). Earth holds water against the siding, which can cause rot, rusting of construction hardware, and mold growth and can encourage termites.

The Plan for Sitework Maintenance

1. Your neighborhood
   a. Review the "Quick Scan" section.
   b. Study the neighborhood as instructed.
   c. Get a sense of soil stability and how water flows in your neighborhood.
   d. Do your checklist.
   e. If you have any concerns or questions, consult an expert immediately.
   f. Discuss how to get neighborhood repairs and maintenance taken care of with the expert.

2. Drainage
   a. Review the "Quick Scan" section.
   b. Study the neighborhood and your parcel as instructed.
   c. Walk your own site.
   d. Get a sense of how water flows off your lot.
   e. If you find any of the following conditions, consult an expert immediately:

Obvious earth movement, continually damp soils, or standing water—anything suspicious about your site can mean big trouble. Don't hesitate. Call an architect or someone with a broad knowledge of sitework and buildings from your "Who Does What Checklist." If the problems look serious, make sure the architect brings in a soils engineer and an earthwork contractor. Listen to all three opinions. Next, call in a general contractor to bring all the findings together and request estimates of repair costs. If you have any doubts about their findings, call in other soils specialists for more opinions. This is expensive repair work, and the more carefully it is analyzed and planned, the better chance you have of getting a good job.

(1) Water does not appear to drain off the lot.
(2) Areas of the earth remain damp.
(3) Trickles or streams flow from underground.
(4) Water is ponding.
(5) Other suspicious situations exist.

f. Do your checklist.
g. Discuss regular maintenance with the expert after any problems are fixed.

3. Soil stability
   a. Walk the building for the Quick Scan.
   b. Get a sense of how earth has moved or settled on the lot.
   c. If there is earth built up against the building, low spots, movement of concrete, streets, or retaining walls, call the expert.
   d. Do your checklist.
   e. Use your expert's advice for setting up the maintenance of the repaired areas.
4. Review your checklists.

Types of Building Sites

Before we move to the "Quick Scan" and on to preparing the checklists for your specific lot, you must know a bit about the way neighborhoods drain and the slope of building sites. First let's take a look at the various types of parcels.

FIGURE 2.3

A flat building site. Most flat sites have some contour and drop to them, as seen in the photo (the lot drops slightly from left to right). The slight rises and falls of flat lots should facilitate draining water away from the structures. However, note how the water flowing down toward the right may flood the foundation of the homes to the right.

The two basic types of building sites are flat and sloped (see Figs. 2.3 and 2.4). Flat sites are the most common. Obviously, they are land that is fairly level, just as the name implies. However, all well-built sites have some grade (slope) to them for removing water from the property.

Sloped lots can include any lot that is not flat. If the grade is steep, the homes are frequently termed *hillside.* They are the result of building in substantially contoured or mountainous locations.

FIGURE 2.4

Hillside building site. Often, if a site is overgrown, it is difficult to see all the contours without a diligent exploration of the site.

Hillside lots are often referred to as *upslope* or *downslope* (Figs. 2.5 and 2.6). For most situations, you can tell what type lot you have quickly. Go out to the street and take a look. Do you go uphill from the street to enter your home? If so, it's an upslope site. If you go downhill, then naturally it is a downslope site. If someone uses other terms, or you are confused, be sure to clarify the discussion.

Both upslope and downslope lots can receive large, fast flows of water from nearby acreage, leaking underground drains, groundwater, as well as lots and streets above them. Periodic checkups are very important for hillside lots. If the developers of the neighborhood understood drainage and were diligent in their design and drain installations, chances are your home is in good keeping with regard to water. However, in some cases developers are not thorough, which can cause hefty repairs for you: Water can back up on streets because the storm drains are insufficient; poor compaction of lots can trigger structural damage; and retaining walls can fail. No matter the quality of the developer's work, you must keep an eye on your site and the lots around you.

FIGURE 2.5

The planter retaining wall and the immediate proximity to the street might add confusion to understanding whether this is an upslope or downslope lot. The typical way to tell whether you live up- or downslope is this: If you have to go up from street level to access the primary structure, you own an upslope property; and if you have to go down, it's a downslope.

Get to Know Your Neighborhood

Before you look specifically at your site, let's take a bike ride or a drive around the neighborhood. Go to the highest place in your local area—this may not be an actual hill but may simply be a rise where the land is higher than the surrounding vicinity. In some places, it may appear that no

FIGURE 2.6

This home sits at the top of the lots next to the street with an alley beside it, which can make upslope and downslope confusing. However, the entry sits down from the road above, so it would typically be called a downslope lot.

spot is higher than the others. In this case, there will be some slant to the street and lawns to provide runoff. If you cannot understand the drainage simply by looking at the environs, try watching the water flow during a wet season and/or call in your professional team.

Take some time at the high place in your locale; get out of your vehicle and walk around. At the very top, notice how water works downward through all the waterways in the area—down the hills, slopes, and streets toward ditches, creeks, arroyos, rivers, lakes, and the ocean (Fig. 2.7). Also, look for steep-cut hillsides, major retaining walls, crib wall (stacked concrete loglike retaining walls), and other large banks of earth that could shift and fall onto your lot or cause your home to move.

Relax, don't rush this process. Survey how drainage starts from the highest ground in your immediate neighborhood. Even if everything is relatively flat, there will be slight drops and rises in the earth. Look carefully at how water drains from high points and then moves on, downward into the drainage system. Ask the following questions as you observe:

- Is water being drained onto your site?
- Does water move away from your home readily?
- Are there places where you are not sure?
- Are their potential landslides that may fall on your home?
- Is there retained earth that could move and affect your buildings?

In well-designed, -built, and -maintained neighborhoods, the earth retention and drainage systems are fairly obvious—except where portions are underground. The water that hits the roofs of the buildings should run off into gutters or onto

 ***Call the Doctor*** 

We strongly suggest that you pay an expert from your "Who Does What Checklist" to walk your home, site, and neighborhood while you are developing your maintenance plan and again, to finalize it. A veteran architect or a general contractor highly recommended by close friends can be an excellent choice. Their eyes can catch things in an instant that would be missed without years of learning from trial and error.

Survey how water flows in your neighborhood. You may not find a hill or other area that appears to be an actual high spot. The street drainage may be subtle, or it may be obvious as in this picture. Look closely and you will be able to understand your unique water course. Note how water will drain downward from roofs and the highest land in the local area into the streets, ditches, and storm drains, then on toward creeks, rivers and lakes, or oceans.

concrete at ground level, and then move off the lot, past the yard and driveways, to the drain system. If all is well, it then flows toward neighborhood waterways like storm ditches and drains—which may be above or under the ground—and does not undermine earth-retaining systems in your locale. See Fig. 2.7.

The main concern for you as a building owner is that water not be standing anywhere on your buildings, concrete, paving, or site—above ground or below—and that the earth is stable.

While you work on jump-starting, it is a good idea to look more than once at how the locale near your buildings is stabilized and drains. If you are bringing in professionals, make sure they drive the neighborhood with you—their quick survey can provide you with a whole textbook of information.

Note how this small lot drains down over the walk into the street. Note ponding at the curb, which can mean that the street does not drain well. Watch areas like this closely.

The Quick Scan

You will use the general framework reference below for monitoring the condition of the site during the "Quick Scan." The list sets up the approach and can help keep you on track while you walk the property.

POINTERS

With sites, even the most avid jump-starters must slow down and understand the way the drainage basin works. If you want to go faster, use the "Who Does What Checklist" from Chap. 1 and get help. Decide who you need to hire—a grading contractor or a soils engineer—to evaluate your site's stability and how it drains. Try them out. If you have any doubts about their knowledge, use your principal expert—your architect or builder—to coach you in finding the correct person.

Quick Scan Outline—The Building Site

As as we stated above, you will be looking at three main concerns while checking out your site:

1. *Neighborhood.* Is it affecting your land and buildings?
   - a. Review the building site checklist (near the end of this chapter) before you go out and walk the neighborhood.
   - b. Notice how water flows down from the high areas along the drainage path.
   - c. Study all retaining walls and places where soil could shift.
   - d. If you are at all concerned that anything in the area may be affecting your home, do not hesitate to get help.

POINTERS

Soil movement, soil buildup, continually damp soils or concrete, and ponding are the most serious things to watch for when you are learning about your site.

2. *Drainage*—water flowing off your parcel.
   - a. After leaving the roof and gutter system, water flows away from buildings.
   - b. Water flows toward the street or other designated drainage systems.
   - c. Is there any place where water is ponding?
   - d. Check carefully for any continuously damp soil.
   - e. Do any areas look questionable?
3. *Soils stability*—movement, buildup, and/or erosion of earth.
   - a. Has buildup been cleared from around the bottom of the framing?
   - b. Does the earth slope down and away from the building?
   - c. Is the foundation affected by moving earth?
   - d. Is there shifting earth on the site?

After you understand the basic neighborhood drainage, go out again and explore the following:

- Walk the neighborhood near your home. Note earthen banks, retaining walls, and where the water goes: down the streets, into curbside storm drains, down the slopes. See Fig. 2.7.
- Check out your lot from the street. Observe how it slopes: toward the street, toward your home, toward the neighbor's yard. Watch for any areas

 **Call the Doctor** 

Never wait when you find damp soil during dry times. Underground pipe leaks, groundwater, and broken water mains are some of the many causes of damp spots. They can trigger foundation failures, mold, and other serious problems.

FIGURE 2.9

This bricked-in yard is humped in the middle. The front slopes toward the sidewalk, and the back slopes toward the residence. Watch out for situations like this—note how it can allow water to stand along the foundations in the rear of the yard.

FIGURE 2.10

A side yard slopes from the curb down toward the home. Water could be ponding in the bushes next to the foundation. Call an expert when in doubt.

where the earth is moving: sunken lawn, hillsides sliding, earth buildup against buildings. See Figs. 2.9 and 2.10.

Now, let's scan your lot and get a handle on the conditions so we can use it to do the checklist.

> **POINTERS**
>
> **The building code requires a 6-inch clearance between the earth and the framing and siding of a structure. This is not really a safe distance because it doesn't include a safety margin for buildup of soils and cellulose. Shoot for a 12-inch clearance, and your home will be more secure as the years roll by.**

- Walk your lot. Inspect the way the earth slopes everywhere on the site; take your time and discover how it really works. Look closely for sunken areas or any earth movement.
- While you are walking the entire building site, inspect for low areas where there is obvious or suspected ponding. See Fig. 2.11.
- Walk the perimeter of your building. Search for low areas where water may stand adjacent to the foundation. Until you know what you are doing, take your time. Watch for areas that are damp during dry cycles.
- Walk the perimeter of your building again. Search for soil and/or plant buildup that is within 6 inches of the bottom of the siding or the framing of your buildings. Take care with this first round. It can be tricky to tell where the siding ends and the foundation begins (see Fig. 2.4). You may have to prune bushes, get behind plants, move items that are stacked against buildings, dig some earth away to find the bottom of your siding, or call an expert. See Figs. 2.4 and 2.13.

FIGURE 2.11

Spotting areas with standing water can require subtle observation. Notice the depressed area in the sunlight to the left of the light tree trunks in the flower bed.

Developing the Site Maintenance Checklists

Now that you have begun to get some sense of what soil movement looks like, and the way that your local vicinity and your building site drain, it is time to get things down on paper. First, go back up to the high place in your neighborhood. Take the book and look over the Building Site Checklist while you are actually observing the vicinity. Also review the three main tasks for taking good care of a site:

- *Your neighborhood.* Tour the locality for drainage and soil movement that may affect your parcel of land.
- *Drainage.* Make sure that surplus water flows away from your parcel readily.
- *Soils stability.* Monitor and control the movement, buildup, and/or erosion of earth.

The next step is to complete the "Building Site Checklist." Like driving to the highest place in your vicinity and walking your property, the "Building Site Checklist" is only a start in helping you really know your property. It is best to think of this as an ongoing project, similar to automobile maintenance. In time you will become very aware of any changes that occur on your property or in your neighborhood. Not only will site knowledge be a part of your maintenance system, but also the information can be invaluable when you undertake projects such as adding rooms or remodeling.

Go through all the line items on the lists, and even if you are the most avid jump-starter, take some time for close observation. Make sure that you take full advantage of the "Notes" column on the list—the more thoroughly you understand

FIGURE 2.12

Notice how leaves and other cellulose matter can build up and turn into very rich soil. This condition encourages rot, termites, and mold, and it is frequently overlooked by homeowners.

FIGURE 2.13

During construction, well before any planting or soil buildup—when this photo was taken—there appears to be a healthy rise from the soil to the mud sill (the bottom 2 × 4 with the bolts through it). However, the siding will stop below the mud sill, and when the landscaping is complete, the soil will be higher than it is in this picture. It is not certain that 6 inches of clearance will remain between the wood and the earth. Maintenance will become critical to keep out termites and water—the owner will have to be diligent to protect the frame of the building.

your home, the more successful your maintenance efforts will be.

Retaining walls are also addressed in Chap. 9, “Landscape and Hardscape.” We have grouped pools, spas, and fountains in Chap. 9 with all of the structures out on the site.

As you work on the checklists, you will discover things that concern you but that you just don’t understand. Experts can help you, and by following their advice, you can get all repairs and fixes taken care of and discuss maintenance of their work with the builders and your advisors.

When you have finished with all your chapters, you will use the “Building Site Checklists,” along with all your other lists, to compile the various Home Maintenance Program Checklists, which you will use each year to help protect your home from destructive elements.

 **Call the Doctor** 

If you have any concern at all that the earth is too close to the siding, framing, or any other wooden or metal items anywhere on your site, do not hesitate to call your architect or builder.

POINTERS

When you set up your Building Site Checklist, remember that water intrusion and earth and concrete movement are two problems that cause very expensive damage to buildings. We look at the causes in this chapter, and we get more specific about concrete work in Chap. 3, “Foundations and Cement Work.”

Building Site Checklist

Done	Item	Call a pro	Notes
✓	GENERAL SITE INFORMATION	✓	
	My neighborhood:		
	Found the highest place		
	Drains well		
	Drains toward		
	My site		
	My buildings		
	Other		
	Large neighborhood drains appear to work well		
	Large neighborhood drainage ditches appear to be clean and work well		

Done	Item	Call a pro	Notes
	Large neighborhood retaining walls appear to be stable		
	Experts called for problems with any of the above		
	Other		
	My home:		
	Plants growing within 12 inches of siding		
	Tree roots		
	Distant but growing toward foundations		
	Close, growing toward foundations		
	Growing at foundations		
	Undermining foundations		
	Safe but growing toward flatwork		
	Close, growing toward flatwork		
	Growing under flatwork		
	Undermining walks, driveway, patio		
	Experts called about questions and problems		
	Other		
	MY SITE DRAINAGE INFORMATION		
	Drain systems in place		
	Gutters on buildings		
	Gutters from buildings to street		
	Drain systems for perimeter of buildings		
	Yard drain system		
	Retaining wall drain systems		

Done	Item	Call a pro	Notes
	Experts called about questions and problems		
	Other		
	My lot drains:		
	Toward my buildings		
	Toward the street		
	Toward an on-site drain system		
	Toward a creek or other body of water		
	Toward a retaining wall		
	Water flows off of my site		
	Experts called about questions and problems		
	Other		
	Visible ponding (water collecting on site)		
	In the middle of my yard		
	Near my home		
	Near outbuildings		
	Near neighboring homes		
	Near my driveway and walks		
	Near my foundations		
	Experts called about questions and problems		
	Other		
	Damp spots on site that never dry up		
	Experts called about questions and problems		
	Other		

Done	Item	Call a pro	Notes
	Retaining walls		
	V ditches and swales draining		
	Water standing above retaining walls		
	Water standing at the foot of walls		
	Water seeping from walls		
	Walls cracked or damaged		
	Through-wall drains working		
	Experts called about questions and problems		
	Other		
	Perimeter of building		
	Damp areas near foundation		
	Plants growing on building		
	Plants holding water at siding		
	Sprinklers watering building		
	Sprinklers leaking		
	Algae or mold growing on building or siding		
	Water standing near foundation		
	Water running away from foundation		
	Experts called about questions and problems		
	Other		
	Drain systems flowing		
	Gutters to street		
	For perimeter of foundations		
	Drains in the yard		
	Curbside		

Done	Item	Call a pro	Notes
	Retaining wall drains		
	Experts called about questions and problems		
	Other		
	Septic system		
	Water standing on earth above		
	Damp soils above		
	Signs of leaking		
	Experts called about questions and problems		
	Other		
	SOILS STABILITY		
	My site has		
	Settled areas		
	Sinkholes		
	Soil that is moving		
	Banks that are slipping		
	Earth slipping out from retaining walls		
	Earth built up less than 6 inches below siding (12 inches is better)		
	Plants that need removal		
	Experts called about questions and problems		
	Other		

Bringing It All Back Home

Site care is extremely critical to the long life of your home or any building you may own. Damp-rot, mold and insect infestations, earth and foundation movement, and cracked concrete are among the most expensive repairs that building owners confront. Nothing can guarantee the life of your home, but diligent care can add to its longevity. A simple, ongoing site maintenance program can save you hundreds of thousands of dollars in future repair bills.

> **POINTERS**
>
> **Overwhelmed? Don't worry about it. You are absorbing a lot of information. Just stay with it. Walk the property and look at the lot and structures. Fill in the checklists as well as you can, but never pass over anything. The beauty of the system is that you are not alone; you have a whole list of experts who will come out and help you.**

Be sure to make the survey a habit. Do it three times a year—before, during, and after the rainy season. When you get to know the vicinity runoff pattern, you may be able to recognize problems that have come about with the storm season. Before the dry season ends, you may be able to notice areas that have remained wet and piles of debris that can stop the flow of drain systems. In the wet times you can actually see slow and standing water. You can spot problems that should be fixed during the dry months. If you want to hand maintenance over to someone else, be sure to take all the tours with them, and understand what they see, just in case you have to change professionals in the future, and so you can keep an eye on their work.

There is no guarantee that you will continue to enjoy good drainage just because you have it now. There is also no way to predict if future development in your neighborhood will undermine your current drain system—new homes can surcharge inadequate systems, causing lots to fail.

As you take care of the tasks on your checklists through the years, watch for water that stands anywhere on the site, especially wet spots in dry seasons. Keep an eye out for any type of settlement, anywhere on your site. Wet spots and settlement are immediate signals for calling a person you really trust. Any number of things could be causing these symptoms: broken water lines, broken drain systems, underground water, or naturally occurring slides that were not taken into account by the builders. They are difficult to understand and often require boring, digging, and other exploratory work to find the cause.

Some people own sites that are adjacent to natural and/or man-made bodies of water. These parcels have their own unique characteristics and require the expertise of engineers and contractors familiar with the locale and the situation at hand.

Ocean and river-front properties can be dramatically affected by the movement of the water and watersheds that drain land above them into the large body of water. Artificial bodies of water can be defective; the liner for the bottom of the reservoir can leak, and many other problems can arise.

With all sites, the main things to watch for are moisture that doesn't go away and earth movement. Often these two dangers can be recognized by events on your lot. For example, concrete lifting or breaking can be a strong indicator of earth moving. If you spot any of these complex situations, call in an expert immediately.

Hills, slopes, ditches, creeks, arroyos, rivers, lakes, or oceans are always a concern. No matter what your situation, while you are walking the neighborhood near your home (do it in the rain with your family!), note where the water goes: down the streets; into curbside storm drains; down the slopes, through your yard, the surrounding lots. The entire neighborhood could be undermining your foundation or causing mold and fungus, which can trigger all sorts of allergies, discomfort, and even serious illness for your family.

One of the most important results of developing a maintenance system is that you will become more and more familiar with your home and its environment. And, the process is similar to becoming familiar with the sounds your car makes—any strange noise, and you go to the mechanic.

The Building Site Resource Directory

Earthwork and site work and many other construction associations. An in-depth list of groups related to arcane subjects can be found. http://www.constructionweblinks.com/Organizations

Excellent information from the National Parks Service. The "What and When to Repair" section is about the foundation and drainage system. Fine material for those who want to learn more. http://www2.cr.nps.gov/tps/roofdown/drainage.htm

A good look at different types of hillside drains that can be used to control the flow of water. http://www.ecy.wa.gov/programs/sea/pubs/95-107/intro.html

Free match to prescreened, customer-rated contractors for all types of jobs! Soils engineer contractors are in your area and interested in your job! We have not used this group, but if you can't find local word-of-mouth information, you might try them and visit jobs that the recommended contractor has done in your area. http://www.servicemagic.com

Informative article about inspecting your site. Check it out. Site drainage issues—how to inspect the grounds of your old house for drainage questions.
http://www.oldhouseweb.com/stories/Detailed/10286.shtml

Interesting, modern plastic pipe system for removing large amounts of water from sites. Gives the homeowner a look at drain systems.
http://varicore.com

The Earthwork Directories at the BuildFind Building Industry Exchange. One of the largest collections of earthwork listings on the Internet. Check them out, especially if you have complex needs.
http://www.building.org/texis/db/bix/+IwwrmwmDeyqww/community.html

CHAPTER 3

Foundations and Cement Work

The vast majority of today's built structures rely on cement for a stable base on the earth where they sit. Mixed with water and additives, cement has changed architecture and buildings forever. The ability to form shapes with ease gives humans a lot of design freedom in construction that was never experienced in the past—the ease of building forms and then pouring concrete into them is a far cry from the labor-intensive task of carving boulders.

Study the drawing in Fig. 3.1, from the Building Science Corporation, which is an excellent cross section of the lower part of a home. Note how the footing is at the bottom with gravel beds that drain water away from the foundation to the street or another system that is designed to handle it. Above the drain bed there is soil. At the top, the earth is graded away from the building at a 5 percent slope. Then, well above grade, the wood siding begins. Figure 3.2 illustrates water stains at the baseboard area. This can be a definite signal that water is not flowing away from the foundation. Be suspicious; these small signs can be ignored easily. But they often indicate that major repairs will mount up.

Imagine that the earth extends away from the building in all directions and creates your lot. If it has ponds in it or if underground water is flowing from the streets or anywhere else, then water can be running toward your foundation beneath the surface, saturating the ground below your home, and weakening its stability.

Look at Fig. 3.1 again and imagine that the water that has not dried up keeps the earth soft. Your foundation supports the tremendous weight of the whole building, which rests on it. The earth supports the

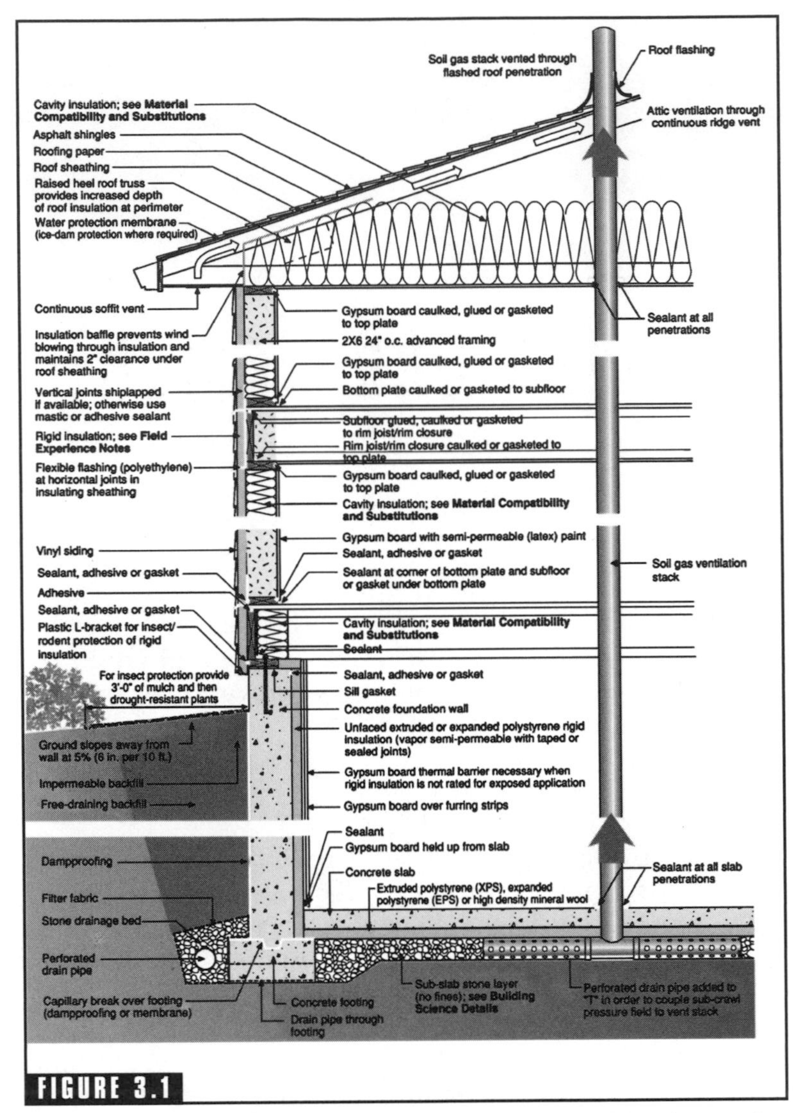

FIGURE 3.1

A modern, well-designed cross section of how a home should work. Note that this structure has a basement and is excellent for cold weather. Use the drawing for basic understanding and ask your experts to fill you in on details about your home. *(Courtesy of Building Science Corporation, © 2003. Reprinted with their permission. For a closer look at the drawing go to the Building Science Web site: www.buildingscience.com/housesthatwork/cold/chicago.htm.)*

whole business, including your foundation, which is very heavy. It can settle, shift, and crack if it is supported by soggy earth.

This chapter gives you another opportunity to explore the material in Chap. 2—buildup of site material against the foundation (see Fig. 3.4). In Fig. 3.1 obviously, if the grade away from the building is altered during landscaping and becomes a level flower bed or, even worse, develops a

Damage at carpet can signal buildup against foundations. It could also indicate damage to a foundation. Keep a close watch for signs of damage inside your buildings. Note the discolored stain on the baseboard outside the tack strip. You can see those stains while the carpet is still in place.

reversed grade in, toward the building, then moisture will be trapped at the bottom of the framing. This situation begins to spread rot, and mold and termites are spawned by the continual presence of moisture. The area adjacent to your foundations must be maintained with continual diligence. (See Fig. 3.5.)

A fine building is very vulnerable to how it is anchored to the earth. When the house outlasts its foundation, repair or replacement can be extremely complicated

POINTERS

The framing and siding of the building must be supported well above the earth. Check your foundations regularly for buildup. Try to get a foot or so—more is always better for keeping your home protected from termites, mold, and rot.

and very expensive. Furthermore, all parts of the structure can be severely damaged by foundation failures. (See Figs. 3.3 and 3.7.)

FIGURE 3.3

Bad structural crack indicates serious concrete problems—call in the pros immediately.

Jump-Starting

Developing the "Foundations and Cement Work Checklist" overlaps with the maintenance requirements in Chap. 2, "The Building Site." Because a building is a living, breathing whole, many parts overlap just as they do with our bodies. The site and all the concrete work are very closely entwined with the foundation—they can all influence the other parts significantly. Studying the components more than once actually boosts your learning curve. Bear with the repetition, because this is very important material and the closer you pay attention now, the better your long-term maintenance plan will serve you.

There are three ongoing tasks involved with the safekeeping of foundations and flatwork (the name commonly used in the construction industry for the concrete areas on the ground around your site—sidewalks, driveways, patios) (Fig. 3.8):

- *Drainage.* Make sure that surplus water flows away from your land readily. (See Figs. 2.8 and 2.10.)
- *Soils stability.* Monitor and control the movement, buildup, and/or erosion of earth.
- *Tree roots.* Survey for roots that are already disturbing your foundations and flatwork.

Below, the plan breaks down the material for compiling your individual "Foundations and Cement Work Checklist"—take a close look at it.

FIGURE 3.4

During your Quick Scan, walk the foundation, looking for all buildup and wet spots around the perimeter of your buildings. Note the rot at the bottom of the wood in the photo. Damage or continuing dampness indicates that you must get an expert immediately.

Next we will walk the lot and look at your foundations and concrete work with the "Quick Scan" section.

The Plan for Foundations and Cement Work Maintenance

Exterior

1. Review the "Quick Scan" section.
2. Walk your foundations and all flatwork.
3. Look for cracks or settlement wherever the foundation is exposed.
4. Notice any places where the foundation or the flatwork stays wet.
5. Hunt for any buckling or leaning of the foundations and flatwork.
6. Search out any areas where the foundations or the flatwork is lifting or settling.

FIGURE 3.5

Water trapped in wood at foundations can lead to serious rot and termite infestation. Note the area (right) that opens to the earth. This area can drain water into the ground leading to settlement of the soil under the concrete step, the stone patio, and, more serious, the foundation of the home. Get all areas like this sealed with concrete, and have the wood at the bottom of the building separated from earth and patios by a foot or so if you want to keep it safe. After problems are corrected, keep water running off the flatwork and all joints caulked. What to do in this situation and developing the maintenance plan will most probably require input from an expert.

Interior

In the House

1. Look for settling floors or any indications that the building is moving.
2. Check out the symptoms in Fig. 3.7, and search the building for any that may be present.

Basement and/or Crawl Space

1. Scan the floor frame for any settlement or moving of the building.
2. Survey all piers and posts for damage.
3. Study the foundation and/or basement walls for cracking or movement.

There is a good chance that you will not be able to see more of your foundation than a bit of concrete just below the siding. The important thing to look for is whether the parts of your foundation that you can see are solid and are not moving, cracked, or sinking.

The condition of having a good deal of the foundation mostly buried and therefore, not visible, can feel like a serious problem to beginners. But don't get bogged down; just walk the building and fill out your checklist. Naturally, if your concern continues, get an expert.

If you have more questions and need more information at this time, go to the "Bringing It All Back Home" section, below and read about the types of foundations and flatwork. This is a very important chapter—fixing concrete work is very expensive; so relax and study hard. As you continue to maintain your home, you will gather more and more knowledge.

Call the Doctor 

For the average homeowner, the main thing is to walk the yard and look for movement or cracking; this does not require in-depth foundation knowledge. But if you see anything that that you don't understand but looks suspicious, this is definitely a time you need help, so get out your "Who Does What Checklist."

FIGURE 3.6

Flatwork drains into garage door creating a potential for considerable damage from water intrusion. Any area around your buildings where the grade obviously propels sheeting water toward the structure requires careful and continual attention. Hire a drainage expert to examine these conditions and fix any problems, then keep on top of the required maintenance.

The Quick Scan

Before you get out and walk the site, look at the list below for a quick understanding of what you will be looking at during the foundations and cement work Quick Scan. Just after the list, you will see the details.

Quick Scan Outline—Foundations and Cement Work

The outline is designed to help keep you focused while you explore your home. You will be looking at three main areas of concern during the survey of the foundations:

1. Movement
   a. Serious cracks
   b. Separation
   c. Rising or falling concrete
   d. Tree disturbance
   e. Anything suspicious

> **POINTERS**
>
> **If the checklist project is confusing or starting to look really hard, remember that the majority of the details cover a lot of information for any first timer. All things in life tend to become easier each time you do them. Also, keep in mind that this is brand new territory—home maintenance systems are not common practice, so you are at the head of the pack. Keep going and get a handle on your home. The system makes diligent maintenance much simpler and much less expensive than the typical approach of just getting random repairs done when there is a problem.**

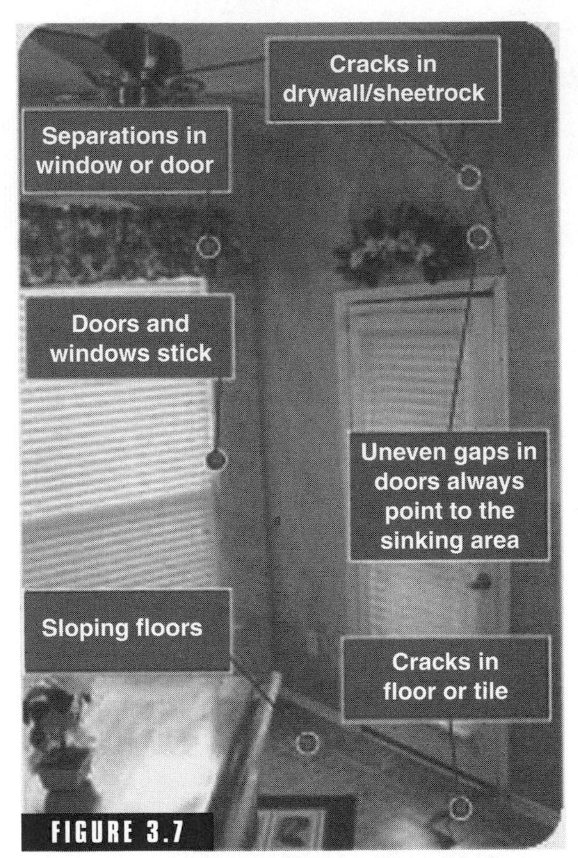

FIGURE 3.7

A look at symptoms of earth and/or concrete problems. This photo is courtesy of Saber Concrete & Leveling Solutions. They fix these problems all the time and their Web site has a wealth of information: *http://saberleveling.com*.

2. Potential movement
   a. Water standing around foundation
   b. Movement of soil at foundation
   c. Movement of soil anywhere on lot
   d. Failure of embankments
   e. Movement of hardscape structures such as swimming pools
   f. Ponding or other moisture anywhere on site

   If you have a basement:
   g. Movement of basement
   h. Movement of walls, framing, or other parts of the basement
   i. Movement of any exposed or other soil in the basement
3. Moisture
   a. Dampness anywhere on the foundation or in the basement
   b. Mineral salt deposits on the foundation or anywhere
   c. Rising or falling of earth or structure
   d. Tree disturbance
   e. Anything suspicious

   If you have a basement:
   f. Damp spots on soil
   g. Movement of walls or other parts of the basement
   h. Movement of any exposed soil in the basement

There are many conditions that cause movement of foundations and cement work. The situations are far too numerous to list in this book, but use the examples to help spot areas around your home. Take a look at the reference section, and visit sites if you want to learn more about foundations and cement work.

For this inspection you will look for defective construction, have it repaired, and begin to get a sense of what you are going to include in your ongoing maintenance plan.

POINTERS

Although cement is very strong, it can fracture just as our bones can; it can scale and break down in the presence of certain chemicals; and water can pass through concrete mixes.

Shifting and cracking flatwork. Note the way the upper slab is raised at the corner where it joins the main slab. It would require testing to actually know; but this probably indicates settlement of the raised slab, up above the drain. This is the type of condition that requires an expert unless you are very good at site drainage. The drain in the slab may be leaking, and damp soils can encourage concrete movement. Note that this can indicate danger to nearby foundations.

As you walk the site, remember that all cracks in concrete are very important. Sometimes hairline cracks are not of consequence, but this is not always the case. In time you can get to know your site and recognize serious cracking quickly, but you should really know your stuff if you are evaluating their importance yourself.

Separations and rising and falling are something to watch for—tree roots or other causes trigger obvious movement of both cement work and foundations. When foundations or flatwork are obviously moving, there is a good chance that something serious such as earth movement is disturbing the concrete work.

Take a careful look at Fig. 3.8 and all the illustrations as you begin to walk the property. Don't rush. Look for the types of situations that are illustrated throughout the book: water standing at foundation cracks, movement of soil, earthbanks mov-

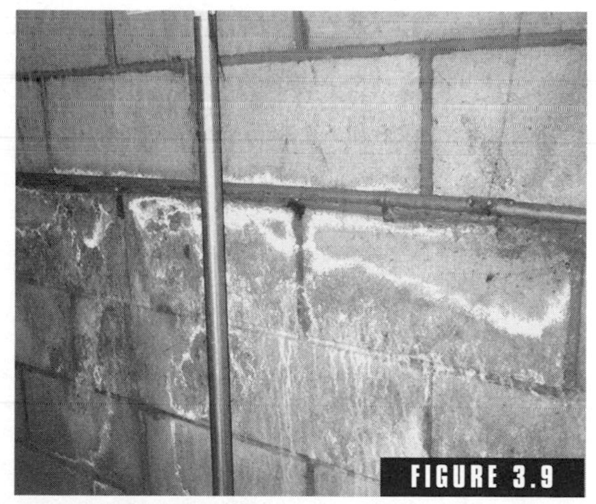

Efflorescence (white deposits of mineral salts on wall) can indicate drain problems. Typically, if any of your underground walls have deposits of white material on them, there is a strong chance that there are moisture problems behind the wall.

 **Call the Doctor** 

Cracks and settlement can have serious consequences on any building and the flatwork outside. Call an expert right away.

POINTERS

Suspicious? Go for it! Trust your intuition! Call your experts! If your misgivings were correct, have the problems repaired and get the builder and your expert to tell you how to maintain their work.

POINTERS

The Foundations and Cement Work Checklist is simple. There is not much to maintaining concrete products. Most of the damages related to it arise from other problems such as site drainage. What you are looking for is cracked and/or moving units that look as if they might injure your buildings or your site. You will have them repaired. Then, with your year-round maintenance plan, continue to monitor them.

ing, hardscape such as swimming pools moving. If you have a basement or a crawl space beneath your home, look closely for cracking and movement near the foundation, walls, framing, or soils.

Moisture must be spotted now, and you must watch for it every year. Dampness anywhere on the foundation or in the basement can indicate underground water or plumbing problems. Check out Fig. 3.9 for an example of mineral salt deposits on a block wall below the earth. These deposits and mold may indicate a defective situation. Be aware of any rising or falling cement, tree disturbances, damp spots, soil movement, or anything suspicious under your home.

Developing the Foundations and Cement Work Checklist

The overlying principle at work when you are thinking about the maintenance of the concrete work is to make sure that the cement construction is stable. Driveways, sidewalks, and patios are all simple to examine. Almost all the problems associated with them are obvious to the untrained eye after just a bit of study; cracking, moving, and spalling (scaly deterioration) can stick out like sore thumbs.

FIGURE 3.10

Yard and patio create ponding. Note the low spot just off the patio. Water will pond in this area for sure. It will cause rot to the posts, and there is a good chance of the patio settling and cracking with time.

The most difficult part of understanding foundation and cement maintenance is the determination of whether the concrete that may be hidden is defective. But don't take chances with what is out of sight—get out your "Who Does What Checklist" and call some help.

Go through all the line items on the lists, and even the most serious jump-starter should take some time for close study. Make sure that you use the "Notes" column on the list—the more completely you understand your home, the more successful your maintenance efforts will be.

Remember that part of the goal in working through the checklists in the individual chapters is finding defective situations on your structure. Do not hesitate to call for advice when you are suspicious of what you discover. If you unearth damage or problems with the current construction, get them fixed. Then use the builder and your experts to set up a maintenance program for their work.

Retaining walls are treated in Chap. 9, "Landscape and Hardscape." We have also grouped pools and spas with hardscape to keep the list divisions simple for newcomers to construction. And now let's move on to building a concrete checklist before the jump-starters go crazy with all this builder babble.

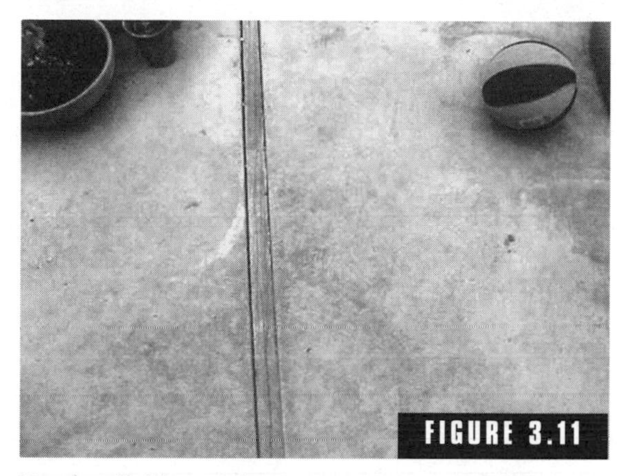

FIGURE 3.11

Wooden dividers rot in time. Note that anywhere they are used in flatwork water can get in under the slab as the 2 × 4s rot.

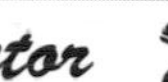

If you are in doubt about anything on any of the checklists, remember that every line item is a crucial part of the home, where expensive repairs can be triggered.

The Foundations and Cement Work Checklist

Done	Item	Call a pro	Notes
✓	FOUNDATIONS AND CEMENT WORK	✓	
	EXTERIOR		
	TYPE OF FOUNDATION		
	T footing—with framed floors		

Done	Item	Call a pro	Notes
	Footing with slab		
	Piers		
	Concrete block		
	Can't tell		
	Other		
	HOUSE SLAB		
	Minor perimeter cracks		
	Deep perimeter cracks		
	Slab stays damp		
	Slab moving		
	Roots threatening		
	Vegetation growing next to		
	Can't see enough		
	Mineral salt buildup		
	Other		
	HOUSE FOUNDATION		
	Minor perimeter cracks		
	Major perimeter cracks		
	Foundation stays damp		
	Foundation moving		
	Roots threatening		
	Vegetation growing next to		
	Gutters draining toward		
	Ponding at foundations		
	Can't see enough		
	Mineral salt buildup		
	Other		
	INTERIOR		

Done	Item	Call a pro	Notes
	SLABS		
	Floors cracked from slab problems		
	Floors moving from slab problems		
	FRAMED FLOORS		
	Finish floors affected by foundation		
	Water stains at base of walls		
	Signs of termite infestation		
	Dampness on interior of crawl space		
	Efflorescence		
	Deterioration of carpet and/or tack strips		
	Hairline concrete cracking		
	Major concrete cracking		
	Can't see enough		
	Mineral salt buildup		
	Other		
	CHECKED CRAWL SPACE		
	CHECKED BASEMENT		
	OTHER		

Bringing It All Back Home

Concrete is the basic structural contact between a building and the earth. I have actually cut open buildings in my lifetime that sat on slices from first-growth redwood trees, using them as a foundation, in direct contact with the earth. The ancient redwood served as long-lived pier blocks. The old trees were very weather-resistant because of chemicals in them, but even extremely resistant lumber can decay over time. Even early con-

FIGURE 3.12

Dangerous mix of flashings. This is the type of situation that will probably be concealed on your home. Note how the metal flashing (metal strip just above concrete—see arrows) can hold water behind it, against the framing if the building paper and siding are not applied correctly and the building is not well maintained.

FIGURE 3.13

Patio sections sinking. Varying top levels of a patio can cause serious trip and fall accidents and involve complex lawsuits. If installation is well designed and executed, and maintenance is provided yearly, tearing up and repouring of slabs can be prevented.

crete would leach and deteriorate and crack; but now it can be designed, mixed, and poured so it produces very long-lived foundations, sidewalks, and driveways.

FIGURE 3.14

Water in planters requires attention. If moisture is trapped where walls are framed with wood, serious rot and termite and mold infestations can easily occur.

Obviously, the foundations are among the most important aspects of any building. They must be built and maintained so they are solid and not threatened by water. Like the site, they must be observed and maintained on a regular basis. Without attention, they can shift, crack, and promote fungal growth, water damage, and termites. They can have a tremendous effect on the framing and interior finishes of your home and require very expensive repairs. If not taken care of, foundations can cause a tremendous loss of property value.

In Chap. 2, "The Building Site," we reviewed how your building relates to the land upon which it sits. If the site is not working correctly with the structures it supports, one of the first places where the problems are evident is the concrete work.

Concrete is typically the basic structural material used in building when contact with the earth is required. It includes your retaining walls, foundations, sidewalks, and driveways. Once you understand Chap. 2, you will have a basic knowledge of where water should be flowing and how the different levels of soil should relate to the various objects around your yard and home that are built out of concrete; and that information carries on over into this chapter.

FIGURE 3.15

Separated lip at pool. Swimming pools require continual attention because of the many vulnerable areas that are continually exposed to water. Note how the walkway has settled to a plane below that of the main patio (gray flatwork).

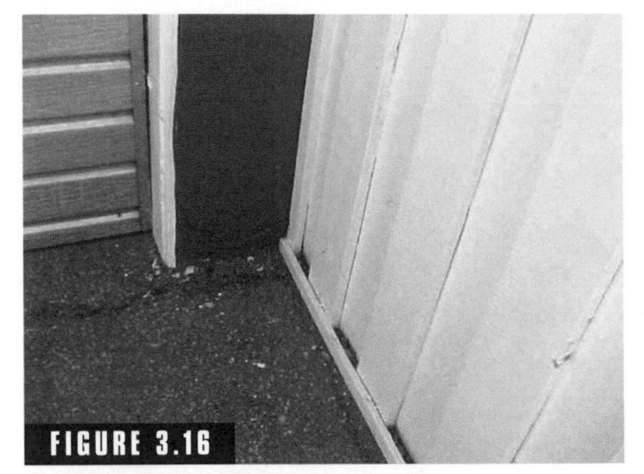

FIGURE 3.16

Foundations—keep them above grade. Wood and siding appear to be buried in asphalt. Note the asphalt crack and the sunken area where the white wood wall meets the dark wall. Water appears to be causing settlement (the garage door area is lower than the main part of the driveway) and probably damp rot to the frame. It may be necessary to raise the foundation height. Look closely for areas like this; they require an expert.

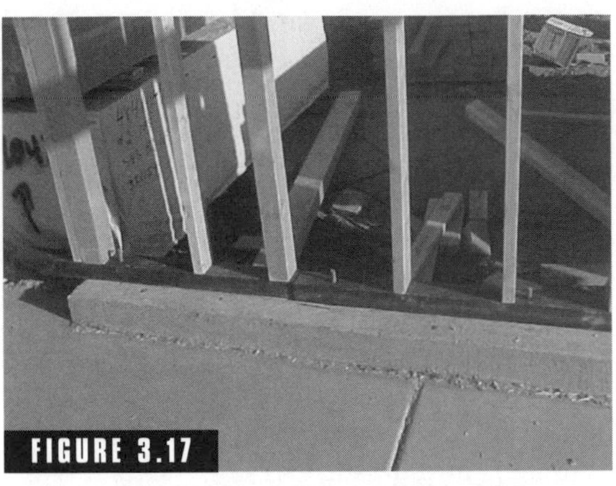

FIGURE 3.17

Clean raised foundation butts to flatwork—note the concrete at the bottom of the framing lumber. This is the foundation, which you may be able to see on the exterior of your building. This is where you are looking for cracks, separation, and movement.

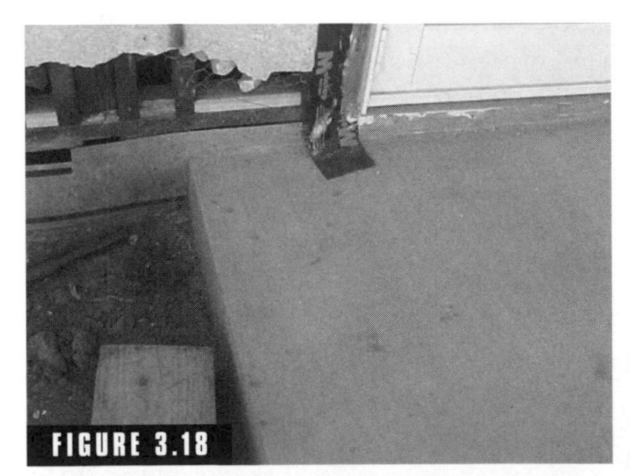

FIGURE 3.18

Foundations—flashings at doors and areas where new slabs are poured against existing walls are very prone to water leaks. All joints exposed to weather must stay caulked.

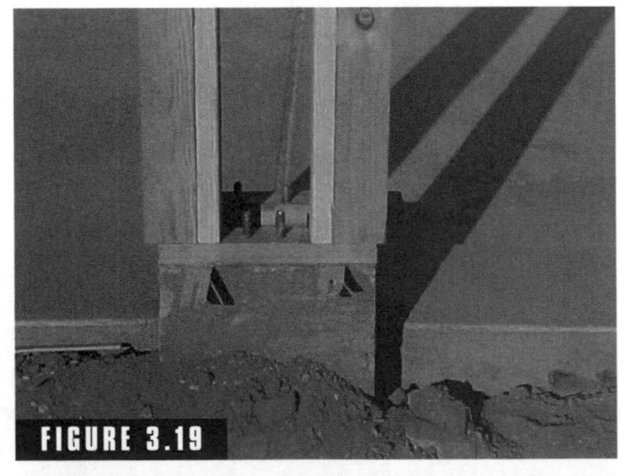

FIGURE 3.19

Raised post base shows a jump in the foundation that is prone to cracking if it is not well built. Note that soil has built up at the garage slab even though the house is not yet completed. This illustrates how soils can build up next to the framing and siding, continuing after construction is complete.

Types of Cement Work

Typically, getting familiar with cement flatwork is pretty simple. Most of it is readily accessible on the site and does not require much expertise to understand. You simply walk your site systematically and look for cracks, movement, and separation of the various flat concrete installations.

When examining flatwork, you want to see that your driveway, walks, and patios are stable, not moving, and in good general condition. Once the experts have visited and problems have been repaired, you must take good care of all the concrete, regularly.

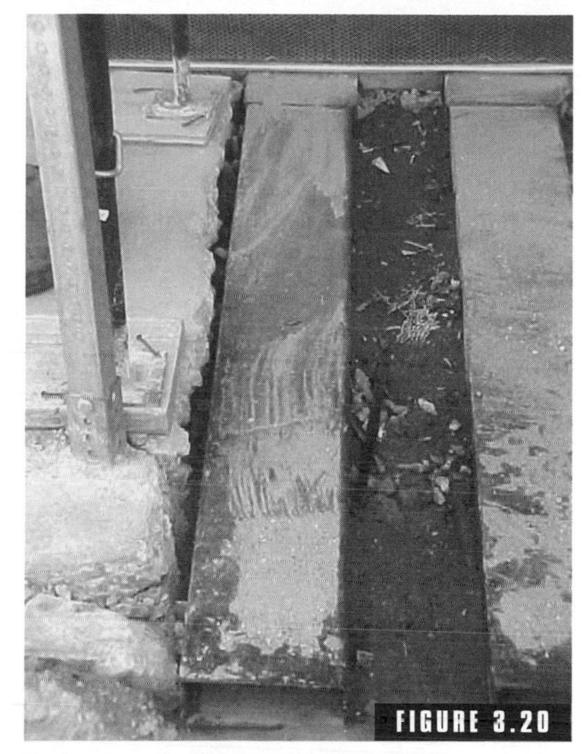

FIGURE 3.20

Drains through walk from building. During remodeling these drains are being installed to remove gutter water to the street, which keeps it from standing on the site.

Foundations

Let's take a quick glance at the various types of flatwork and foundations. There are a few things you need to be aware of that may help you improve the checklists for your specific lot—let's take a look.

As we stated at the beginning of the chapter, the bulk of foundations installed today are variations on a simple T footing. Take a look at Fig. 3.1 again, and you will see that the main stem of the foundation looks like an inverted "T." Basement foundations (Fig. 3.1), crawl space foundations, monolithic slabs, and slabs on grade are a number of basic variations on the theme.

There are also many other foundations—concrete block, ballast trench, piers, and grade beams—but the scope of this book does not permit the study of all of parts of a home. If you wish to learn more about the various types of foundations, check the Resource Directory at the end of this chapter and search concrete and foundation books online. You will find an abundance of information.

Foundations are one of the most important aspects of any building. They must be built and maintained so they are solid and not threatened by water. Like the site, they must be observed and cared for on a regular basis. Without attention, they can shift, crack, and promote fungal growth, water damage, and termites. They can have a tremendous effect on the framing of your home and require very expensive

FIGURE 3.21

Post raised off grade. Note that the concrete footing is now covered with stucco and the frame sits well above water, but the trim piece at the bottom of the stucco is not caulked and water will sit against the framing lumber. Keep all separations at joints between the siding, stucco, or masonry veneers and trim caulked tightly.

FIGURE 3.22

Stucco at concrete is holding water—this will cause rot in the building frame, and draw termites and produce mold. Where the foundation area of the building joins with any concrete flatwork such as at walks and patios, there must typically be a 2-inch gap between the concrete and the siding. Make sure to do whatever it takes to keep the framing dry. You may need an expert for help with this. Note white line on the wall, between the dry area and the damp part—mineral salt deposits from water standing, then drying on surface repeatedly.

FIGURE 3.23

This penetration is between the patio (below) and the stucco (above) at the foundation. It is a serious situation because water will be soaking the framing for long periods of time. This can cause rot and draws termites, promoting extensive hidden damage. Note the caulk separated from the joint—to the right—in this photo. If caulk is being used to seal a siding-to-concrete joint, the concrete is probably poured too high. With it this close to the framing, a design professional is required to orchestrate the repair. As you walk your foundation, note any conditions where caulk is not sealing a joint, where metal is exposed, or where there are holes in the siding. These situations require an expert, as well as repairs, before maintenance procedures can be established.

repairs. They can cause a significant loss of property value so you must always make them one of the maintenance priorities.

The Foundations and Cement Work Resource Directory

About the Concrete Foundations Association. Associations can sometimes be a help in finding contractors. http://www.cfawalls.org

Basement do-it-yourself waterproofing indoors with Sani-tred. This is a liquid rubber sealing method if you find leaks in the basement or crawl space. Discuss it with your expert. http://www.sanitred.com/BasementWaterproofing.htm?ref=findwh

Concrete block foundations. Block can make a good basement. Some good basic information about blocks. http://www. atlasblock.com

Concrete foundations. Some good foundation information.
http://www.quikrete.com

Concrete foundations as the finished floor. Very interesting if you face major work.
http://www.greenbuilder.com

Concrete Homes Council. Lots of resources.
http://www.concretehomescouncil.org

Concrete Homes Home Page. More, excellent information.
http://www.concretehomes.com

Concrete Network. Information about countertops, staining, and stamping.
http://www.concretenetwork.com

Concrete repair and resurfacing kits. Patching and repair kits for concrete floors.
http://www.interstateproducts.com/concrete.htm

Concrete Secrets video. Learn about doing your own concrete work.
http://www.concretesecrets.com

Foundation repair. Learn about methods of foundation repair. Find local contractors.
http://www.saberpiering.com

Frost-protected shallow foundations. Excellent cold-weather ideas.
http://oikos.com

Need to clean or seal? Learn more about your hard surfaces.
http://aldonchem.com/

Perma-Crete. Resurfacing of old concrete.
http://permacrete.com

Radon mitigation and waterproofing concrete sealer. Find out about radon and mold.
http://www.radonseal.com

WarmlyYours.com. Put your heating system right in the concrete.
http://www.warmlyyours.com/

Wet basements. Basic information if you have a basement.
http://www.drybasements.com

CHAPTER 4

Framing

The frame of a building is something like the skeleton of a body, and the roof and siding resemble the skin. Although damages are mostly hidden from view, many factors can harm a building's framing, and because walls must be opened up (Figs. 4.1 and 4.2) repairs are very costly. There are a number of types of frames: stone, concrete block, log, timber, steel, etc. But during the twentieth century, stick (2×4 lumber) frames became far and away the most common in the United States.

Stick framing is vulnerable to motion, water, and living creatures. It can suffer from many problems like moisture intrusion, which causes it to rot; termites and boring insects; the site shifting; earthquakes; wind; and foundation and other movement. It is extremely vulnerable to the condition of the siding, which can allow water to drain into the building and cause it to warp, move, and rot.

The other types of skeletons, for example, stone, block, timber, steel, and rammed earth, are not specifically covered here. The material is too extensive, and the subject is maintenance, not a study of framing. However, this is not a problem because the majority of a building's frame is covered and the various maintenance requirements are similar for the protection of any type of skeleton.

Naturally, you want to bring in a professional from your Who Does What Checklist if you have any questions about the safety of your structural and decorative framing. For example, odd-looking areas beneath the finishes on exterior beams that support patio covers or decks can indicate problems. If you poke an ice pick into the frame of

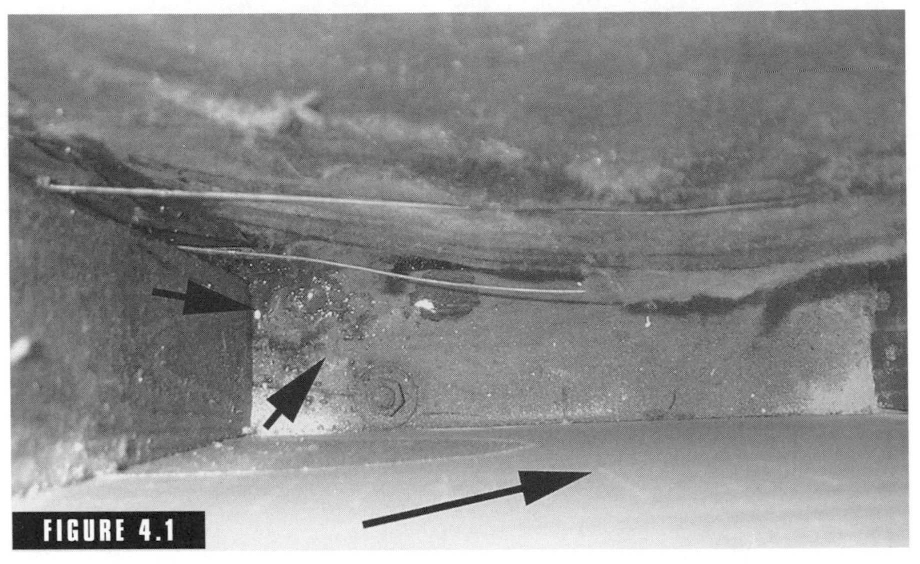

FIGURE 4.1

Water leak seen in wall—rusting j-bolt and nut. You are looking down the sheet rock (bottom white area where right-hand arrow is) and down a stud (where the left-hand arrow is) to water spots (left arrow points at one and center arrow, resting on bottom mudsill points at two) on the bottom plate next to a rusted j-bolt. This shot, looking down inside a wall, illustrates how vulnerable your home is to water leaking in, then rotting the lumber, spreading mold, or drawing termites.

FIGURE 4.2

Inside your wall frame with black building paper exposed. Note that the black felt (arrows are pointing at it) is all that protects the frame of your building beneath the siding, stucco, or brick. In most homes more than 5 years old, it is just paper impregnated with tar, and if it suffers a long period of exposure to water, it disentegrates.

the structures and the wood is spongy, the problems can be very serious.

Since the framing is very important and must be tight and dry (Fig. 4.2), but is mostly invisible to the building owner, the reader will welcome insights into how it is best observed and maintained. This chapter explains how the frame works, where it can be observed, and when specialists should be consulted. It includes the various areas of a building: roofs, foundations, crawl spaces, attics, and the perimeter of the roof.

Jump-Starting

The framing of a building requires regular attention, but the maintenance is fairly simple. There are two main tasks involved with keeping the frame safe from rot and organisms, and they both involve other parts of the building.

FIGURE 4.3

Side window paper must lap (arrow is pointing at lap) over bottom paper. Lapping is very important. If the side piece is under the strip below the window and the caulk is not maintained at the window frame, water will keep the frame wet all winter. If the caulk at the window is maintained with care, it can protect the frame against lapping defects beneath the siding.

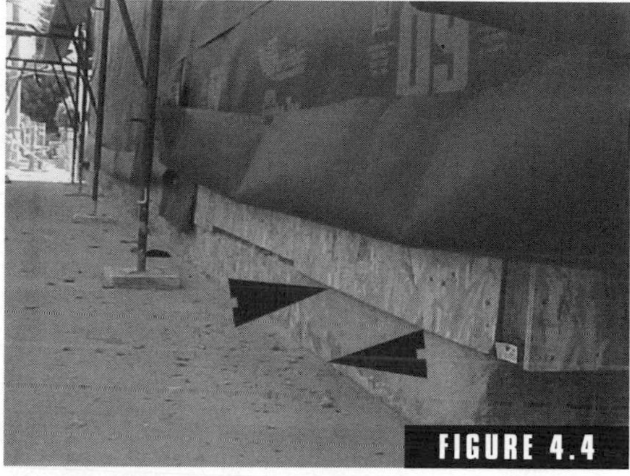

FIGURE 4.4

Framing set well above grade (arrows point to top and bottom of concrete). Note the large space above the flatwork. Unless the windows or roof are not well designed, built, and maintained, this building can always stay nice and dry. With proper maintenance it will be solid for a very long time.

- *Shedding water.* The roof and siding must be tight. All joints and penetrations must be solid and well caulked, and paint must be in tip-top condition.
- *Earth movement and buildup.* Keep all soils and plantings carefully tended so they do not hold water against the bottom part of the framing.

Next we look at the plan, which will direct you through putting your own framing checklist together. When we get to quick scan, we will go outside again and take a look at the framing wherever we can see it. Then we go back inside, to climb up into attics. Then, down under the building to the crawl spaces.

The Plan for Framing Maintenance

1. Exterior
   - a. Review the "Quick Scan" section.
   - b. Walk your site.
   - c. Get a sense of how the siding covers the frame of your home.
   - d. Notice any places where the frame is visible.
   - e. Study all trellis, patio cover, and other framing that is an exterior appendage to the house's frame.

FIGURE 4.5

Modern framing on new slab. Note that the framing (mudsill—top right and middle arrows point to it) being placed on the slab is well above grade (the earth line—bottom left arrow—is at grade), but it will soon be reworked with grading equipment. Then a landscaper will come in, and finish grading will take place. In a very short time plantings and earth will build up and there may be earth holding water against the siding.

- **f.** Look for any exterior damage to your framing.
- **g.** Start your checklist.

2. Interior
- **a.** Attic
  - **(1)** Scan the underside of the roofing material for water leaks.
  - **(2)** Check out the rafters for any water damage.
- **b.** Basement and/or crawl space
  - **(1)** Scan the underside of the flooring material for water leaks and mold.
  - **(2)** Examine the floor joists for any water damage.
  - **(3)** Survey all piers and posts for damage.
  - **(4)** Look for mold and fungus in all locations.

3. Interior and exterior
- **a.** Though you need help when you have any questions, if you find any of the following conditions, get your expert immediately:

(1) Holes in siding expose framing
(2) Rot found on any exterior framing such as patio covers
(3) Rot found in attic, basement, or crawl space
(4) Moisture stains leaking through interior walls
(5) Termites found anywhere
(6) Cracked or moving foundations affecting framing
(7) Other suspicious situations

b. Discuss regular maintenance with the expert after any problems are fixed.

 **Call the Doctor** 

If you do not know whether you have termite flashing, call your architect or builder right away and evaluate the solution for your home.

 **Call the Doctor** 

If tight spaces cramp your style, you doubt whether you will understand what you are looking at, or you just plain old feel dumb crawling around in the spider webs, it's expert time. While the professional is there, start filling in your checklist.

Types of Framing

Framing of our structures began with people using products located close to the building site. Humans have tried many materials for the skeleton of their buildings—caves, poles, logs, staves, and wattle. The primary form today is the stick frame (2 × 4 lumber) home. Steel and block are also used as well as some more exotic products that are attempts at environmental balance: straw bale, rubber tire, and rammed earth. Check these out on the Internet for some very interesting reading.

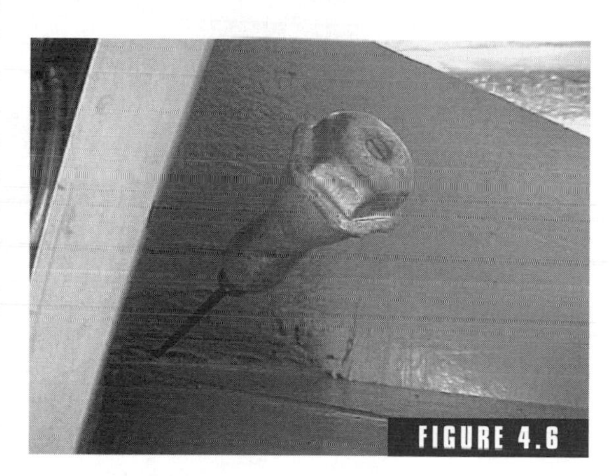

FIGURE 4.6

Use an ice pick to check for rot. When you are doing your framing inspection to put your maintenance plan together, a common ice pick is excellent for testing lumber for deterioration.

In Fig. 4.8, you can see that wood frames are very vulnerable to rot and termites and that unseen damage may be spreading in your buildings as you read this. Steel frames are also prone to degradation from water intrusion—in Fig. 4.8 rust is the primary symptom at door jambs if the studs and mudsill are metal. However, note that no matter what the frame of the building is built from—even something as stable as concrete block—termites and rot can cause major damage to interior framing (Fig. 4.10), doors, and windows, and that mold can grow almost anywhere that stays damp.

No matter what type of frame, the two basic maintenance needs run true: keep the penetrations—including doors and windows—caulked,

FIGURE 4.7

Serious cracks may signal structural problems. They must be caulked until repaired. After the cracks are fixed, include the maintenance for the new surface in your maintenance checklist and be diligent.

FIGURE 4.8

Framing is very vulnerable at doors. Any signs of water stains on interior walls can indicate damage to framing lumber beneath the sheetrock. One of the primary benefits of a maintenance plan is knowing that the caulk around door and window openings is in good shape.

and keep earth buildup down at the base of the building and water draining away. Even if the building is concrete block, numerous elements can affect the lower areas and damage the doors and windows—care is a must for any structure.

The Quick Scan

The next step is to start exploring the chassis of your building and creating the "Framing Checklist." Use the list below to keep yourself focused during the survey.

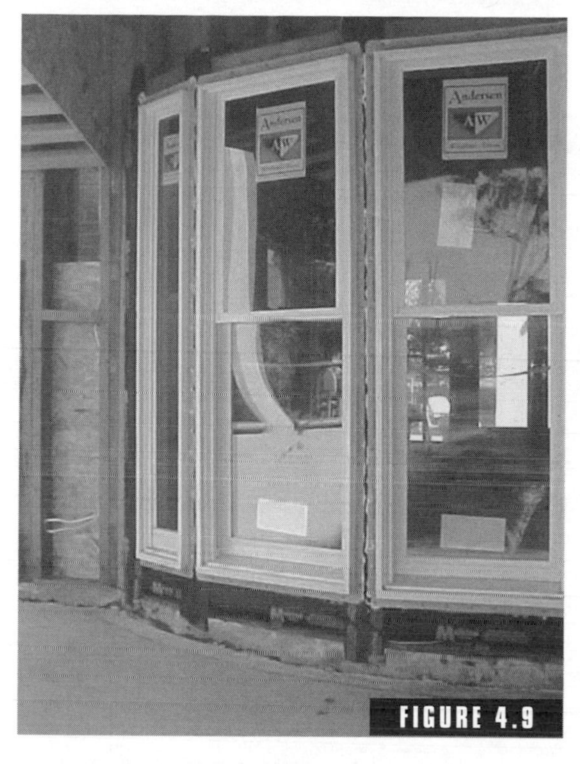

FIGURE 4.9

Curved walls can lead to severe rot. Framing for bay windows and other complex wall patterns can leave small areas of siding, stucco, or brick that are prone to movement—note the narrow areas between the window frames. The caulk must be watched with extra diligence in these areas in order to avoid leaks.

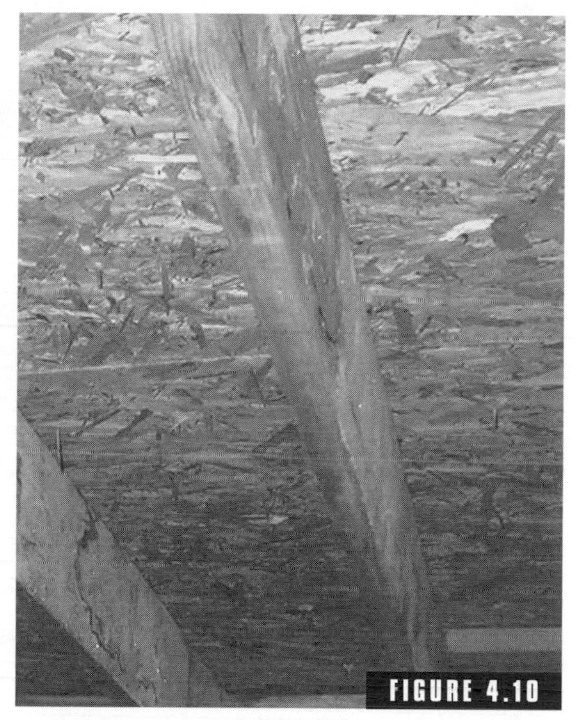

FIGURE 4.10

Even in new homes, poor ventilation can grow mold quickly. Check all the crawl spaces carefully for growths. Mold can spread very heavily, and it can be bad for a family's health. Call in an expert if you find any growth or suspect it in inaccessible areas.

Quick Scan Outline—Framing

You will be looking at three main concepts while hunting for framing defects:

1. Shedding water (This material is so important that it is covered again in both Chap. 5, "Doors and Windows" and Chap. 6, "Siding.")
   - a. All siding is in place.
   - b. Penetrations in siding are sealed. (We include stone, brick, stucco, and any exterior wall surface as siding.)
   - c. Exposed framing.
2. Earth movement and buildup
   - a. Buildup cleared from around the bottom of the framing.
   - b. Earth sloped down and away from the building.
3. Exterior framing

FIGURE 4.11

The frame above a window. The large timber is a header. There is a strong member above all door and window openings because studs that supported the weight above have been removed for the window hole. Any leaks seen in this area are typically from the floor above or the roof.

FIGURE 4.12

Transitions at bottom of wall threaten framing—flashing can be out of sync and damage can be caused by outside forces. Watch these areas with care.

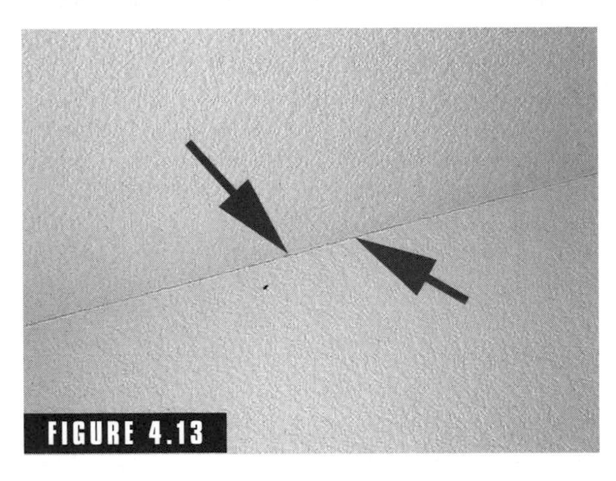

FIGURE 4.13

Ceiling cracks may indicate framing problems. They can be due to structural defects in the engineered wood and metal connections located in the upper part of your home, or they can be indicative of framing damage.

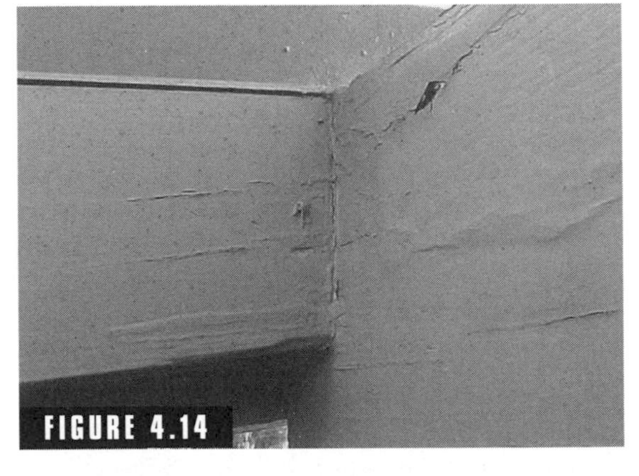

FIGURE 4.14

While you are walking your property, be sure to take close note of any exterior framing that may have problems, whether you can actually recognize them or not. As well as the obvious rot holes, note the bulges in the paint. Also, remember that the breaks in the paint surface allow water to get into the space between the paint and the wood. The paint then performs just the opposite function from which it was intended—it traps moisture, promoting rot, rather than causing it to run off the wood.

a. Areas where water may be trapped between framing members.
b. Are all appendages and exterior framing made of products that are resistant to the element's product or well painted?
c. Watch for water trapped in joints of exterior trellises and decks.

4. Other.

If You Have a Crawl Space in the Attic

1. Water stains on the roof frame
2. Termite infestations
3. Mold growth
4. Signs of rot
5. Damp spots
6. Separated framing joints
7. Water stains on the roof sheathing
8. Othor

If You Have a Crawl Space or Basement below Ground

1. Water stains on the floor frame and sheathing
2. Water stains on the walls or wall framing
3. Termite infestations
4. Mold growth
5. Signs of rot
6. Damp spots
7. Separated framing joints
8. Missing posts
9. Other

FIGURE 4.15

Relation of window framing to roof framing. Note how easy it is for roof leaks to migrate down to the window framing. The wood with the speckled pattern is called *strand board*. It is used as structural sheathing to hold the framing in the building stable. If it rots or comes loose from the nails, the entire building is vulnerable to shift, racking, and numerous defective conditions.

POINTERS

Even the most durable frames, such as concrete block, require diligent maintenance. All frames can contribute to moisture problems. Make certain that your experts know your type of frame very well—if it is a steel frame, your professionals must have experience in steel frame construction.

> **POINTERS**
>
> **Keep a close eye out for buildup of soil and plants next to the perimeter of your buildings at ground level. Both conditions trap water against your siding and are often the cause of a great deal of unseen damage—a 6-inch space is code. But try to keep 12 inches between your siding and grade if possible.**

When you get out on your lot, you really start to see how much of the framing is hidden. As we have said, this makes it difficult to know if you have horrible damage or not. We urge you not to take a cavalier attitude with this part of your home—framing damage is among the most expensive of the repairs that building owners ever face. And many framing problems can be discovered, opened, fixed, and maintained before repair costs have ballooned into the six-figure realm. If anything you find during the Quick Scan is suspicious, don't take chances, get some help.

Let's take a look—be sure to remember that most of the framing will be covered with building paper and siding. You are looking for indications that moisture may be getting into the frame of your building—it is very important to you because water trapped next to structural members is one of the prime causes of major home repairs.

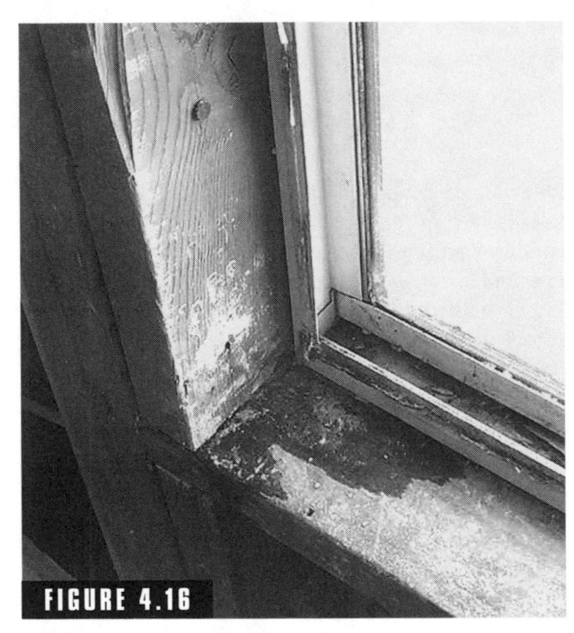

FIGURE 4.16

Water stain on framing can indicate that termites are being attracted to the moisture, the finishes are being severely damaged, mold is being attracted, and rot will soon begin.

> **POINTERS**
>
> **Any mold, termite activity, or rot noticed on the framing is urgent. Get help right away.**

AT THE EARTH LEVEL

- Walk the building again, as you did for the earlier chapters. This time, think about what it looks like inside the walls at the bottom of the framing. If you are not able to visualize, go back in this chapter and look at the photos. Imagine the hidden parts of your buildings.
- If earth is built up against the wall at the mudsill or if there are any holes in the siding, even up by the windows, nothing but the felt-impregnated building papers protect your framing—and it rots within months of direct exposure to water.

ON THE SIDING

As you walk the perimeter, take a close look at the walls. Notice all penetrations, corners of the building, trim boards, any place that looks as though water could enter through the skin of the building and get trapped next to your framing.

At the Door and Window Level

Notice things like the crack in Fig. 4.7. Watch for caulk at the door and window frames. Figure 4.16 illustrates how water begins to pond and launch the many things that can happen to the framing if door and window openings are not maintained.

Notice the caulk remaining on the sliding door frame in Fig. 4.8. The siding has been removed, and it is very obvious that water entering at this joint is a big threat to your estate's checkbook. It is important to maintain the bead of caulk between the door frame and the siding at every door and window, all over your buildings. See Chap. 5, "Doors and Windows," for more about this vulnerable part of your home.

 *Call the Doctor* 

Crawl spaces and attics can be extremely hard to access, and they are very important areas for finding defects and monitoring the effectiveness of your maintenance program. An expert may be required.

 *Call the Doctor* 

If you are in doubt about items on any of the checklists, remember that every listing is a crucial part of the home, where expensive repairs can be triggered.

Higher up on the Building

As you walk, stop and take a look at all the levels: foundation line, door and window line, and then, while you pause, check out the roof line. Pay close attention to the edge of the roof and to any structures such as decks and patio tops or trellises that are add-ons to your building. Take a look at the photos in Chap. 7, "Roofs," and any other photos that help you. Gutters and other add-ons often trap water and cause severe rot—a knowledgeable expert will see the danger signs quickly.

On the Inside

Back inside, get down in the crawl space and up into the attic areas, and look for defects and signs of water intrusion.

Developing the Framing Checklist

A lot of the Framing Checklist work takes place by inference, because so much of it is hidden inside the walls of your building. With structural framing, the object is mainly to look at exterior and interior indications that there may be defective framing or damage to structural components from water intrusion. If there are signs of damage, calling an expert is important. Much of the deterioration caused by exposure to water can be a good deal less expensive to repair if you catch it early. This is similar to dental problems, where it is better to get a filling than a root canal.

The Framing Checklist

Done	Item	Call a pro	Notes
	FRAMING		
	EXTERIOR		
	TYPE OF FRAMING		
	Wooden 2 × 4		
	Wooden timber		
	Light-gauge steel		
	Steel beam		
	Concrete block		
	Other		
	GENERAL		
	Trellis attached to building		
	Patio cover attached to building		
	Deck attached to building		
	Other structures attached to building		
	Odd wall joints on exterior of building		
	Siding is solid		
	Frame visible through siding		
	Other		
	EARTH AND PLANTINGS		
	Inspect for termite guards—get them installed in all possible places		
	The earth is 6 inches below the framing		

Done	Item	Call a pro	Notes
	EARTH AND PLANTINGS		
	The earth next to the bottom of the framing drains away from the house		
	Plants are pruned away from the bottom of the building		
	Bushes are pruned away from the siding		
	Trees are pruned away from the roof line		
	Other		
	SIGNS OF LOWER WALL FRAME DAMAGE		
	Lower-level framing at house needs repair		
	Building needs repair from attachment of lower-level exterior framing		
	Siding affected by exterior framing		
	Doors and windows affected by exterior framing		
	Damp siding and/or framing continual		
	Termite evidence visible		
	Rot visible		
	Other		
	SIGNS OF UPPER WALL FRAME DAMAGE		
	Upper level framing at house needs repair		

Done	Item	Call a pro	Notes
	Building needs repair from attachment of upper-level exterior framing		
	Trim board behind gutters affected by exterior framing		
	Soffits affected by exterior framing		
	Roof affected by exterior framing		
	Damage to boards in gutter areas		
	Rot visible		
	Other		
	INTERIOR		
	BASEMENTS AND CRAWL SPACES		
	Sagging framing		
	Supports appear to be missing		
	Cracked lumber		
	Damaged framing		
	Rusted framing metal		
	Rot at framing lumber		
	Mold and fungus		
	Efflorescence (mineral salt deposits) on wood or concrete		
	Signs of running water on lumber		
	Damp wood or concrete		
	ATTICS		
	Signs of roof frame damage		
	Damp or damaged underside of roofing materials		

Done	Item	Call a pro	Notes
	Daylight visible through roof		
	Damp framing materials		
	Signs of moisture on lumber, metals, and other items		
	Mold and/or fungus growth		
	Sagging roof frame		
	Other		

FIGURE 4.17

Cracks at places such as door casings can indicate soils, structure, or moisture problems with the frame of the building. Study the interior with care. If you suspect that there are problems, never hesitate to ask for help. If repairs are required, set up the maintenance with your architect and builders.

FIGURE 4.18

Modern garage with metal in framing is vulnerable to rust. The galvanized surfaces used to deter rust are only minimal deterrents—most will degenerate when exposed to mineral salts and water. They must be kept dry or painted with rust-resistant paint. This is true on the roof deck or wherever galvanized metal is used.

Bringing It All Back Home

The framing maintenance system is similar to plumbing, electric, and HVAC because the majority of the building's frame is invisible. Most of the upkeep that is needed will be closely related to other parts of your maintenance program: foundations and cement work, doors and windows, siding, roofs, and landscaping. Care to these other areas of your buildings can do an awful lot to protect the frame of your structures.

This is extremely important because the repair costs for the frame are right up there with roofs and foundations, and sometimes they are worse because you will be forced to relocate while the contractor is working on the problems.

Termite repair companies are very often the biggest unforeseen cost when a family puts a home on the market. But in many situations if the principals had been privy to the material in this book for study of the property, fixing whatever problems they found, setting up the maintenance checklist system, and providing diligent attention yearly, the repair costs for most homes would have been cut dramatically.

The Framing Resource Directory

Balloon frame houses. This is a fun history of where our wood framing came from.
http://www.uh.edu

Framing the house. More pictures—how your frame works.
http://www.thumbknitting.com

Framing walls and trusses, Prairie House, and Hometime. Can give you a sense of how complicated repairs can be.
http://www.hometime.com

Healthy house framing. Check into steel framing and healthy homes.
http://www.toxicfreeliving.com

House framing glossary. Good glossary—learn the terms.
http://alsnetbiz.com

Howstuffworks, "How House Construction Works." Excellent information about how homes are framed.
http://www.howstuffworks.com

*Timber Framers Guild.*If you have a timber-framed home or are simply interested, see this site.
http://www.tfguild.org

Wood-frame houses: Introduction. Very good information about wood-framed homes. The seismic information is not just limited to California—there are earthquake fault lines all over the United States, the whole globe.
http://www.johnmartin.com

CHAPTER 5

Doors and Windows

No matter what they are made of doors, windows, and the openings in which they sit are extremely vulnerable to the elements on a continual basis. Their weep holes clog, not allowing water to drain away; they warp, rot, decompose; and water slips in around the jambs, causing serious damage. (See Fig. 5.1.)

This chapter explains how to keep a close eye on your doors and windows. The checklist covers the many things to watch out for: shrinking, warping, rotting, hinge movement, slider track leaks, drainage at window seats, gaps at siding, and finishes. The list is long.

Doors and windows are a must for every building, and they get some of the most severe physical use that any part of a structure receives. There's not much to maintaining interior doors; just a mild dose of attention when they bind or squeak will keep them in good shape. Exterior doors and windows need to be watched continuously, caulked, repainted, and maintained, but the care can save you and the heirs to your estate hundreds of thousands of dollars in future repair bills. (See Figs. 5.2 and 5.3.)

There are two distinct divisions of doors and windows—those that are exposed to the elements and those that simply receive wear and tear from the heavy use they receive. Moving interior windows are not that common, and typically they get very little use compared to the busy life of a door.

We use this natural pattern of inside and outside windows and doors to help guide you into this chapter. First, we take you outside again,

FIGURE 5.1

Weep holes must not be clogged. You will find the weep holes (they are the round ones in this picture) at the bottoms of frames of windows and some doors. If they are plugged up, water flowing down the glass will drain into the building. Also note that at the corner, the two parts of the frame are separated (the upside down, L-shaped crack). Defective or damaged window frames can allow water into the building.

FIGURE 5.2

Window deep in brick or any surface requires that the joint between window frame and the siding, stucco, brick, or any building covering always stay caulked.

FIGURE 5.3

Crowded windows require particular care. The top piece of moisture barrier (the black strips behind the nailed flanges) is not in place yet—note how vulnerable the moisture barrier is to water if the siding leaks. Most moisture barrier is not a true flashing, and it rots in several months of severe, continual dampness.

walk the house, and look at the exterior units. After studying the outer surfaces, we return to the indoors for a look at the minor maintenance required for interior doors and windows.

 **Call the Doctor** 

If you are not knowledgeable about doors and windows and you see any wet spots or damage that looks suspicious, bring in an expert from the Who Does What Checklist. This one visit can save you many thousands of dollars over the life of your home.

Jump-Starting

The maintenance of doors and windows is straightforward, but remember that they are absolutely one of the most critical parts of your home. Wood needs moisturizing, sealing, or painting (Fig. 5.4). The newest, most weather-resistant window and door products must be maintained according to the factory instructions. Even aluminum will pit and deteriorate if it is not cared for properly.

As with all the checklists, while studying maintenance you will also work toward uncovering what needs to be repaired or fixed with the help of your experts. As the repairs are being done, go ahead with the "Doors and Windows Checklist." Often, working people who are very knowledgeable and can help you understand confusing, defective areas in a jiffy.

Three goals are the primary thrust of superior door and window maintenance:

1. *Protection of the units.* The vast majority of door and window units need to be cleaned and sealed on a regular basis.
2. *Repair and lubrication of working parts.* When doors and windows bind or are hard to work, don't shine it on—and, do not slam them and beat on them. Lubricate them or fix them to keep the problems from getting worse.
3. *Stopping of all water entry immediately.* Doors and windows are secured into the frame of the building with door and window jambs. The jamb is the decorative wood around the door or windows in their frames. The jambs are secured

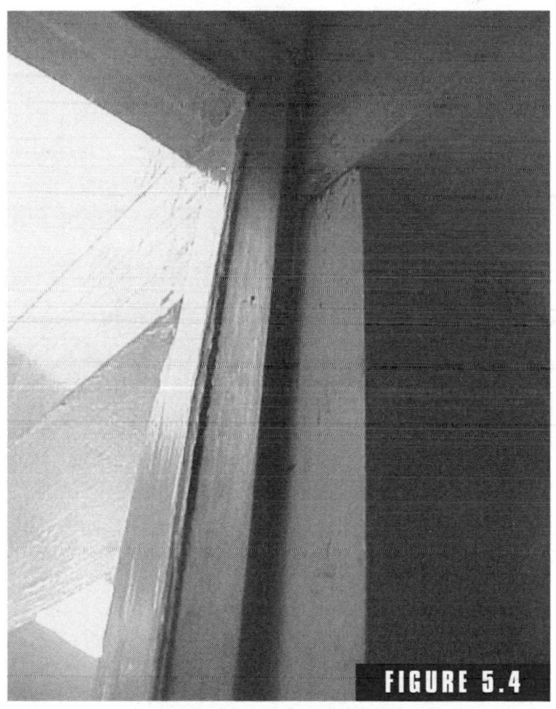

FIGURE 5.4

Wood window frames require care. Frames that are built at the site require especially close attention: Caulk must remain solid, and painted surfaces need to remain tight and uncracked. When water seeps in beneath cracks in exterior paint, it is actually sealed against the wood and severe damage is likely because the wood does not get a regular chance to dry out.

into a rough opening in the frame of the building. *Both* the surround of the door or window in its jamb *and* the crack between the door or window jamb and the siding must be sealed from moisture intrusion at all times.

The following plan serves to clarify how you will break down the information for putting together your specific Doors and Windows Checklist. After you review the list, we will tour your structure and examine the openings more closely.

The Plan for Door and Window Maintenance

Exterior

Doors

1. Check out the condition of all units.
   a. Hard to open and close.
      (1) Call in an expert (unless you are very handy).
      (2) Follow the expert's maintenance advice.
   b. Finishes cracked or peeling.
      (1) Refinish.
      (2) Keep them maintained.
      (3) Any glass sealed properly.
   c. Veneers, glue joints, frame damaged.
      (1) Get advice or replace.
      (2) Keep them maintained per manufacturer's instructions.
   d. Caulk around door jambs is tight.
      (1) Scrape it and replace if needed.
      (2) Keep them maintained.
   e. Damage beside or at bottom of door jamb.
      (1) Repair with professionals unless you are very handy.
      (2) Keep the doors and jambs maintained.

Windows

1. Check out the condition of all units.
   a. Hard to open and close.
      (1) Call in an expert (unless you are very handy).
      (2) Follow the expert's maintenance advice.
   b. Finishes cracked or peeling.
      (1) Refinish.
      (2) Keep them maintained.
      (3) All glazing compounds and sealants in good repair.
   c. Veneers, glue joints, frame damaged.

(1) Get advice or replace.
(2) Keep them maintained per manufacturer's instructions.

d. Caulk around window jambs is tight.
(1) Scrape it and replace if needed.
(2) Keep them maintained.

e. Damage beside or at bottom of window jamb.
(1) Repair with professionals unless you are very handy.
(2) Keep the windows and jambs maintained.

Interior

Doors

1. Check out the condition of all units.
   a. Hard to open and close.
      (1) Call in an expert (unless you are very handy).
      (2) Follow the expert's maintenance advice.
   b. Finishes cracked or peeling.
      (1) Refinish.
      (2) Keep them maintained.
   c. Veneers, glue joints, frame damaged.
      (1) Get advice or replace.
      (2) Keep them maintained per manufacturer's instructions.

Windows

2. Check out the condition of all units (the vast majority of modern homes do not have moving interior windows—pass by this section unless you have interior windows).
   Hard to open and close.
      (1) Call in an expert (unless you are very handy).
      (2) Follow the expert's maintenance advice.

Types of Doors and Windows

We will go outside and take a close look at the doors and windows and then look inside for any evidence that water intrusion has caused damage to the structure already. But before we start walking the building, let's take a look at the various styles of doors and windows.

There are many types of doors, all the way from bank vaults to leaded glass. The purpose of this book is to develop your checklist maintenance system, not to study building materials. For that reason, we limit each survey section to a rapid review of common items related to the particular category. Let's take a look at some common doors and windows.

FIGURE 5.5

Bad delamination of a door. Using a door that is not designed for the situation is never advisable. If a poor grade of door is used, it requires strict attention to maintenance.

Doors

For the reader's convenience and to avoid repetition, doors are separated into two groups, interior and exterior. Several characteristics are common to both; others are unique to each type. For example, rot and water intrusion will very seldom be a consideration in the maintenance of interior doors except shower doors, which can leak and cause considerable damage. (Shower doors are included with plumbing in Chap. 8 checklist.)

Exterior Doors: Typically, in residential construction, hinged exterior doors are either wooden or metal. They may or may not have glass set in them, can be composite with some plastics, and they may be varied in design such as a Dutch door or French doors. The other typical exterior door for residential construction is the sliding track door. Currently, they are most often made with aluminum frames although some come with other types of housings such as wood (Fig. 5.5).

The hinged and sliding doors require four basic maintenance procedures on an ongoing basis:

> **POINTERS**
>
> Wooden exterior doors are extremely vulnerable to the elements. Even the toughest exterior glues can be harmed by severe combinations of water, mineral salts, and sun. Keep all joints filled and painted with thorough diligence.

- Keep the units clean and the paint and other finishes intact.
- Make sure that any drain devices such as weep holes are working.
- The openings around the doors and windows—including the bottom thresholds—must be watertight. (See Fig. 5.6.)
- Make sure the doors open, close, and lock smoothly in the door jambs.

> **POINTERS**
>
> Aluminum door frames can be affected by the elements. Keep them clean and protected with products prescribed by the manufacturer.

Interior Doors: The doors inside your buildings will probably never inspire lawsuits or cause expensive repairs, but they can become loose in the door jambs, the latches wear out, and sliding tracks and pivoting mounts can bind and break

Built-up trim catches water next to window. Note that a 2 × 6 board (center of photo) nailed on the strand board sheathing (left of photo) next to a window (right side of photo) traps water badly if the caulk around the window is not maintained.

down (Fig. 5.7). Although this book is devoted to items that cause extremely costly damage, you can keep an eye on how well your interior doors work and learn to use an appropriate lubricant on latches. Have hinges, tracks, or other binding problems repaired as soon as they pop up.

Exterior Windows: The exterior windows of your home can be made of many different materials: steel, aluminum, plastics, wood, and combinations of these. All the different types require ongoing maintenance, and it is very important that it be carried out on a systematic basis. What we are studying here is how you can set up the checklists to help prevent serious damage to the entire building, and that information is mostly generic. For example, the caulk bead between the siding and the frame of the window must be kept tight. (See Fig. 5.8.)

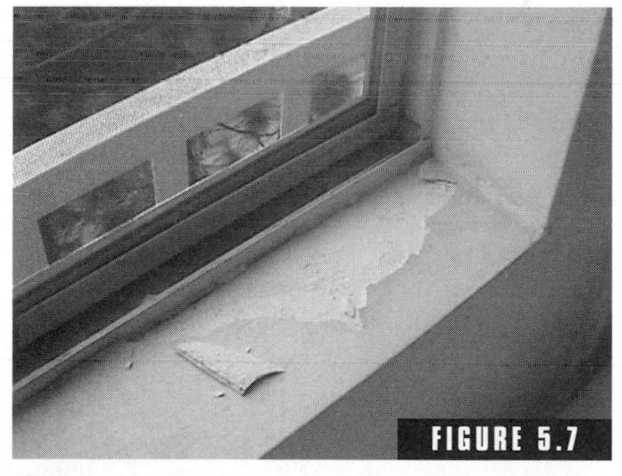

Check inside your home regularly—window seats can be warning signals for rot. Other areas are indicators, too—the wall next to the bottoms of doors, just above the floor line, beneath windows. Many areas can reveal damage from water leaks. Mold and fungus growth can also be spotted on interior walls.

The vast array of windows being used at this time prohibits an in-depth survey, but the information is readily available if you wish to familiarize yourself with your own type of window. If you know the manufacturer—some windows are marked

Call the Doctor 

Finishes must be repaired as soon as you spot damage, and they have to be applied to new doors and windows immediately. Weather can cause damage very quickly. Unless you are really handy, use a professional with a lot of experience to work on door and window finishes. There are many problems that can arise for inexperienced crews. For example, some solvents can damage the sealants that are used to keep moisture out at glass joints.

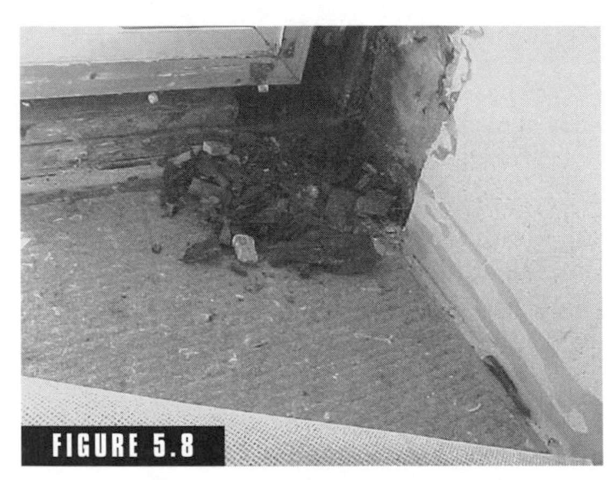

FIGURE 5.8

Lack of caulk can cause severe damage. Notice that if you were looking at the outside of your home, the damage would be at the lower left of the window you were looking at—somewhere above that area something is not properly sealed. Water has cascaded through the leak, and the framing lumber has stayed wet for long periods of time.

with the manufacturer's name—search out the site on the World Wide Web. We have listed some of them in the Reference Directory. Typically exterior doors are either wooden or metal. If you don't know the brand, your experts may be able to figure it out or help you find a product on the Internet that is very similar to your own.

If you don't know the product, you must take good care of your home anyway. Your expert can point out the important items such as weep holes and how to keep the openings in your building for the windows watertight. (See Fig. 5.9.)

There are several basic types of site-installed and factory units: fixed, such as picture windows that are framed and installed on the job, sliding windows, casements, double-hung windows, and variations on the designs. The basic, ongoing maintenance requirements are much the same as those for exterior doors:

- Keep the units clean and the paint and other finishes intact.
- Make sure that they shed water from any draining devices such as weep holes on the window units.
- The joints around every window, whether it moves or not—including the bottom casings—need to stay sealed to prevent water intrusion.

Pointers

Latches, locks, strike plates—all the things that move on doors and windows—should remain unpainted unless they are designed to receive the paint suited for the job and it is applied in an appropriate manner.

FIGURE 5.9

Watch for telltale carpet stains. French doors and sliding glass doors both have wide thresholds. Because the span is wide, it is prone to moving and creating gaps for water to come in from the exterior of the building.

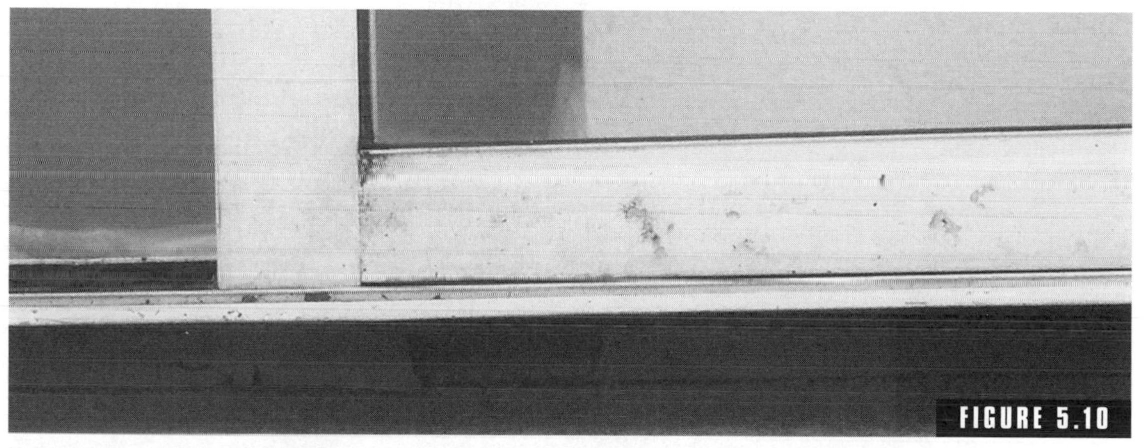

FIGURE 5.10

Vulnerable under the sliding door, the threshold is walked on continually, which causes it to move. The shifting opens gaps through which water can flow during the wet season, and all year if the patio is flooded.

- Make sure they open, close, and lock smoothly in the window casings.

Interior Windows: On the inside of your home, just as with the doors, your moving windows are not typically going to cause extensive damage to the rest of your buildings. In fact, you may not have

> **POINTERS**
>
> **Be careful with glazing compound and putty—some remains soft for several months.**

any moving windows or fixed windows that are entirely within the structure. However, there are exceptions such as transoms—pivoting windows that allow air to move from one part of the space to other parts. Basic upkeep will make the unit last longer; but again, it is doubtful that interior windows will ever lead to expensive, major repair work.

Unless you are very skilled at door and window work, you may want to use your "Who Does What Checklist." Leaks at exterior doors and windows trigger a great many problems and require a person with overview to spot dangerous symptoms. (See Figs. 5.11 and 5.12.)

POINTERS

Cleaning of windows is made easy with a high-quality squeegee. You can save a lot of money over the cost of packaged cleaners, too—pour a quarter cup of rubbing alcohol in a gallon pail, and fill it up with water. Get a soft bristled window brush at the janitor supply, be careful on ladders, wipe the edges with newspapers (they don't leave lint), and you will have spotless windows fast.

FIGURE 5.11

Mudsill rotting under window—exterior caulk must be maintained with great care.

The Quick Scan

Before we put your doors and windows list together, let's go out and take a walk. The Quick Scan Outline will help keep you focalized while reviewing the doors and windows.

Quick Scan Outline—Doors and Windows

There are two things to watch while you are surveying the doors and windows:

1. The connection of the frames of doors and windows to the siding and the earth. (This is one of the most critical joints on a building, causing huge amounts of damage.)
   a. All caulk is in place.
   b. Units are secure in openings.
   c. Earth is not built up to them.
2. Door and window units are solid, sealed, finished tightly, and close weather-tight in their frames.

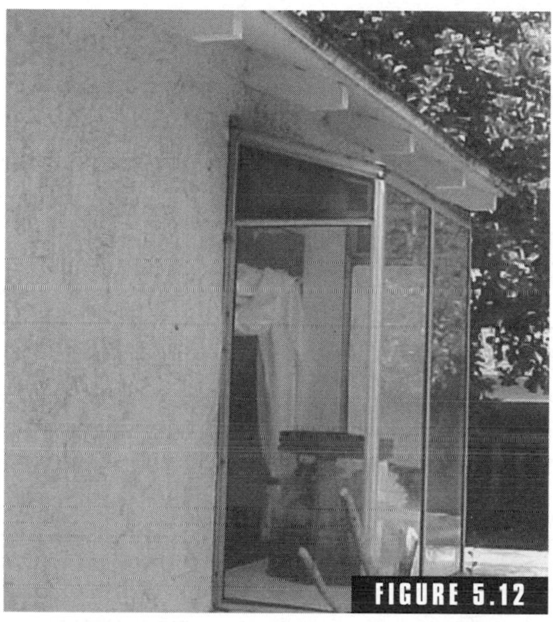

FIGURE 5.12

Pay close attention to vulnerable windows. Add-ons should be watched with care. Note that there is a space between the bottom of the roof line and the atrium window where blown water can easily be trapped and stay damp for long periods of time.

Walk the house and stop at each window and door to examine them closely. The problems are typically quite obvious after you catch on. For example, there has to be a good tight bead of caulk all the way around at the joint from frame to siding.

Many times, with door and window rot, mold, and termites, some evidence of the problems can be seen. For example, holes in the wood from termites or rot. But the majority of any problem will probably be hidden and require opening floors and walls to understand the extent of needed repairs. Just the same, you will probably be able to get some idea that there are problems if they exist. But do not depend on yourself unless you are very well versed in construction repairs.

The following gives you some sense of how to scan the openings and the units.

> **POINTERS**
>
> **If you never maintain anything else on your home, keep water from entering at door and window openings, and around the edges of the frames where they join the siding.**

AROUND THE DOOR AND WINDOW JAMBS

Just as in the earlier chapters, walk and look closely. Study all the frames with care, watching for missing caulk, cracks, or anything that looks suspicious. Be sure to check out the bottoms of all openings. You may need your expert for this inspection. (See Fig. 5.13.)

Door and Window Units

Look closely at the surfaces and where the glass sits in the frame, and make sure that all paint, finishes, sealants, and caulk are solid. Tops, bottoms, and sides of doors and windows are also supposed to be painted. It is highly important that you have a coach for doors and windows.

Developing the Doors and Windows Checklist

The main idea when working with your doors and windows is that you want protective coatings and glue joints to be intact and abrasive elements that will harm them cleaned off the doors and windows. And, you don't want water entering the building anywhere around them.

The Doors and Windows Checklist

Done	Item	Call a pro	Notes
✓	DOORS	✓	
	EXTERIOR		
	TYPES OF DOORS		
	Front door		
	Rear door		
	Side doors		
	Sliding doors		
	French doors		
	Specially designed doors		
	Garage doors		
	Other		
	All doors caulked tightly at joint between door frame and siding		

Done	Item	Call a pro	Notes
	All doors sealed tightly at joint between bottom of door frame and siding or concrete slab		
	All doors sealed tightly at weather stripping		
	All doors sealed correctly between door and threshold		
	All doors sealed tightly at top joint between door frame and siding		
	All doors sealed tightly at top joint between door and door frame		
	All door paint and other protective coatings sealing door tightly		
	Other		
	INTERIOR		
	TYPE OF DOORS		
	Hinged doors		
	Swinging doors		
	Pocket doors		
	Sliding doors		
	Other		
	All doors open and close smoothly		
	All door handles work well		
	All door locks work well		
	Other		
	WINDOWS		
	EXTERIOR		
	TYPE OF WINDOWS		
	Wood		
	Aluminum		
	Plastic/vinyl		

Done	Item	Call a pro	Notes
	Steel		
	Built-up product		
	Fixed windows		
	Sliding windows		
	Casement windows		
	Other		
	All windows caulked tightly at joint between door frame and siding		
	All windows sealed tightly at joint between bottom of frame and siding		
	All opening windows sealed tightly at weather stripping		
	All opening windows sealed/ weather-stripped correctly between window and window seat		
	All windows sealed tightly at top joint between window frame and siding		
	All moving windows sealed/ weather-stripped tightly at top joint between window and window frame		
	All window paint and other protective coatings sealing door tightly		
	Other		
	INTERIOR		
	TYPES OF WINDOWS		
	Wood		
	Aluminum		
	Plastic/vinyl		

Done	Item	Call a pro	Notes
	Steel		
	Built-up product		
	Fixed windows		
	Sliding windows		
	Transom windows		
	Other		
	All windows open and close smoothly		
	All window mechanisms work well		
	All window locks work well		
	Other		

Bringing It All Back Home

Doors and windows are the portal between the occupants of a building and the world outside—doors bring the building to life. They are among the most important systems in any home. They are unique in that they offer a very close relationship between the human activity on the inside and the out-of-doors.

On the exterior, doors tie in closely with several of the other areas that require close attention. For this reason a number of the tasks that are required for doors and windows will be incorporated with a couple of other parts of the building when we build your maintenance program in Chap. 10. Siding, roofs, and landscaping can all have an effect on the door and window categories. The savings on termite work alone, which can be prevented by sound care of doors and windows, can easily roll over the $100,000 mark.

FIGURE 5.13

Water stains at window seat. Another example of the paint peeling away from water getting into the building. Notice on the sheetrock next to the window frame that this one is stained on the sidewall too, which indicates that water is leaking in from higher up, as well as probably entering at the bottom of the window.

FIGURE 5.14

Stains at sliding doors may indicate damage. Note that moisture intrusion has started to have an effect on sheetrock.

A good deal of the integration of the parts of the building is hidden from the site, as you learned while studying the building frame. For example, asphalt-impregnated building paper is supposed to be lapped evenly down the side of your building. At doors and windows, sisal paper surrounds the windows (see Fig. 5.15), and it must be interwoven to lap correctly with the black building paper.

The Doors and Windows Resource Directory

You can learn a lot about doors and windows by visiting manufacturers' Web sites. The sites we have included provide general information about doors and windows along with their product information.

Andersen Corporation. Good illustrations of window parts.
http://www.andersenwindows.com

Best Practices: Window Installation. Flashing a window.
http://www.ibacos.com/pubs/WindowGuidelines.pdf

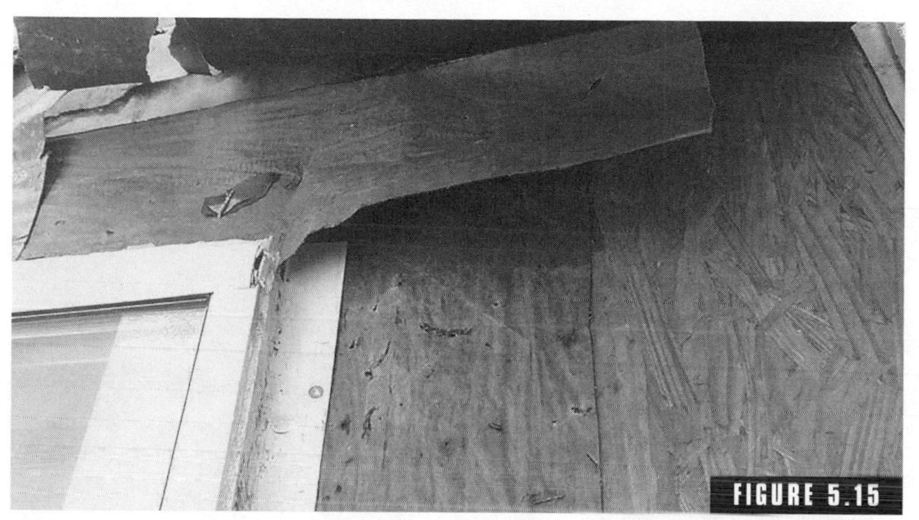

Lapping of building paper at windows. Note that the brown paper (sisal or Kraft) is tucked under the window's nailing flange. Imagine what water would do as it flowed down this intersection—it would flow under the flange (right) and then leak into the frame of the building. All the paper must be lapped correctly. This paper is often called *flashing,* which is not correct. In the UBC (Universal Building Code) it is referred to as a *moisture barrier,* which is a limited function. The paper can rot quickly if it is wet for extended periods of time. If the window to siding joints are not kept caulked, your buildings will most probably suffer from rot and/or termites and mold.

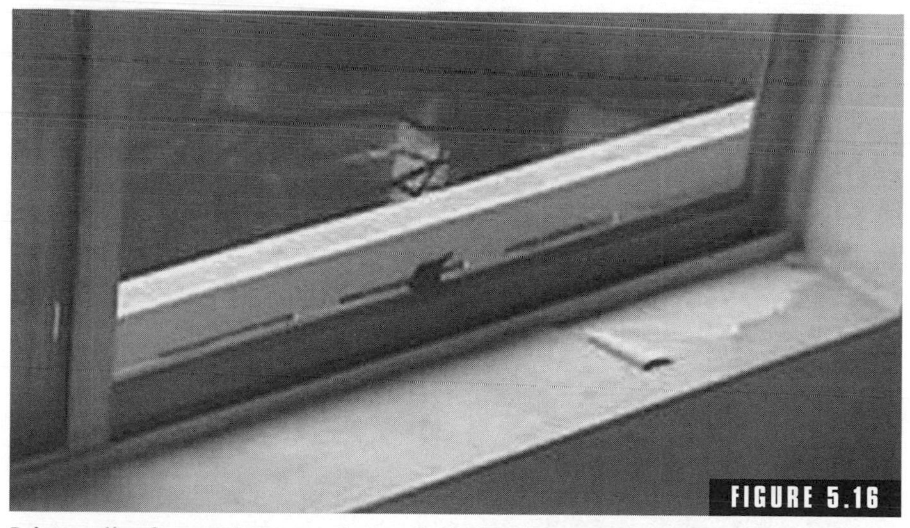

Paint peeling from moisture on sheetrock window seat. Note the exterior railing and the screen. All parts of the building must be considered when defects are discovered. If the exterior railing is attached through the siding, the penetration could be the culprit or at least part of the source of water intrusion. Complications are one of the main reasons for bringing in experts to help put your checklist system together.

FIGURE 5.17

Window is too high for owner to be aware of moisture. Just as with the roof, don't let areas that are not easy to examine slip past maintenance inspections.

FIGURE 5.18

Trim on wood siding must stay caulked. Water sheeting down the side of the building or blown in during a strong rain can move right in under the trim boards at windows and trigger, rot, mold, and termites in the frame.

Construction Defects. A construction lawyer and expert witnesses illustrate problems at doors and windows.
http://www.constructiondefects.com

Handyman USA—doors and windows. Door and window information.
http://www.handymanusa.com

HandymanWire—window leaks, questions and answers. Tight windows.
http://www.handymanwire.com

Integrity Windows and Doors—window and door anatomy. This is an excellent site for learning some of the basics, very well done.
http://www.integritywindows.com

Leaks at windows. More about leaking. This is about a stucco installation, but some of it applies to any siding.
http://www.allaboutstucco.com

Marvin windows and doors—landing. Some good information.
http://www.marvin.com

Pella windows and doors—home. Some tips on maintenance.
http://www.pella.com

Replacement. Information about window replacement.
http://www.myaffiliateprogram.com

Simonton windows—vinyl window and patio door installation instructions. Good information with drawings.
http://www.simonton.com

Stained glass windows and leaded glass doors. Some information and facts about stained and leaded glass.
http://www.glass-by-design.com

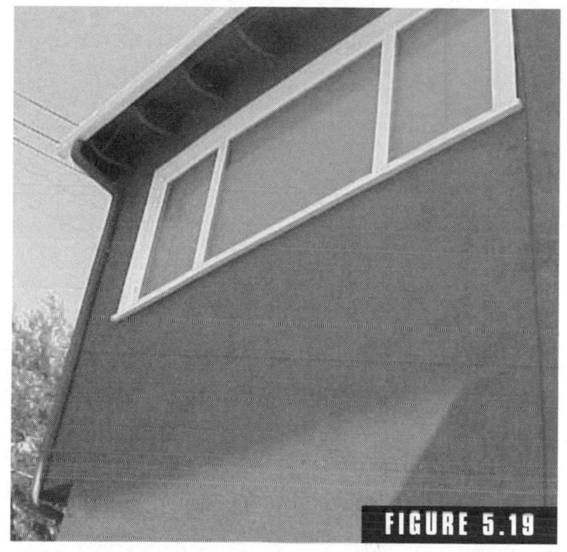

FIGURE 5.19
Even hard-to-reach windows must stay caulked.

Vinyl windows and patio door installation. Information about retrofits.
http://www.dehavenconstruction.com

Water infiltration—Design Build Business Magazine. Water infiltration in the structure of a home is often referred to as a "silent killer." It can be a slow and steady destruction and can go unnoticed for years. Information from the builder's perspective.
http://www.designbuildbusiness.com

Water infiltration and mold remediation. Numerous resources for learning about mold and water leaks in buildings.
http://www.lightlayer.com

Water infiltration testing and repair. Testing for water leaks on buildings.
http://www.gulfcoastinspection.com

Window and door. New product information.
http://www.windowanddoor.net

Window and door installation. Excellent. Check it out.
http://www.leeric.lsu.edu

Windows and doors on the industry window and door site. General information.
http://www.windowanddoor.com

Wooden window. Good information.
http://www.woodenwindow.com

CHAPTER 6

Siding

No matter what the style—stucco, ship-lapped wood, plywood, or any vulnerable material—the sides of your home require regular maintenance to avoid severe damage to the siding material, wood or other trim products, the frame, and even the foundation. Even products such as concrete block, and metal or vinyl coverings, which are relatively stable against the elements, require inspection and upkeep at critical areas such as the window openings and the roofline.

In this chapter the primary goal is to look at the siding as you would the skin on your body. It is an osmotic membrane; it breathes and can accommodate the movement of the frame as it absorbs moisture in the wet season, swelling up, and shrinks during the dry season, as moisture evaporates.

The continual shifting of the building creates cracks at the door and window frames, trim boards, and the corners of the structure. Although the maintenance is not overly demanding or complex, it does require ongoing attention. Let's move right into the preparation for the checklist by looking at the skin of your building more closely.

POINTERS

The skin of the building at the windows is one of the main places where water intrusion takes place—and this triggers errors and omissions lawsuits and construction defect litigation. The whole process of maintenance will be readily paid for if you keep your home so tight that you decrease the chance of needing major repairs. Even if you must pour hours of your time into complex litigation, you are less likely to require major repairs.

FIGURE 6.1

Keep the siding free of cellulose matter. Leaves, plants, and soils that build up within a foot or so of the siding can cause moisture to be trapped in the siding, against the framing materials. This leads to mold growth, rust, and rot in the framing and can cause the siding itself to decompose, as you can see in the photo.

Jump-Starting

As you put together the Siding Checklist, you will need to understand that the maintenance work on the skin of a building is closely integrated to most of the other parts of the structure. At the top, the roof and the siding are closely connected with each other. The same is true at the doors and windows and at the foundation. If there are leaks through the siding, they can have serious effects on the framing and on both the interior and exterior finishes. See Figs. 6.3 and 6.4.

The requirements for conscientious care of the siding on a building will be present every season on your final checklist because the skin is so important. Let's take a look at the essentials of what needs to happen.

- *Shedding water.* Siding must work as a solid membrane that continually deflects water.
- *All openings sealed.* This includes any holes, penetrations for vents, and the joints around doors and windows. Caulk and sealants must filll all voids and be new enough to form a resilient seam.
- *Paint or other membranes always maintained.* No matter what the final surface is, it must never be left to deteriorate, or water will be trapped behind.

The plan for your individual Siding Checklist follows. After you are familiar with it, you will walk the building again and get follow-through with the "Quick Scan" section.

The Plan for Siding Maintenance

1. Review the "Quick Scan" section.
2. Walk the perimeter of the structure.
3. Is the siding snug in place and solid?

Siding damage can indicate more serious problems. Observe the strip of concrete to the left. The earth is built up toward the siding, but there is still space between the earth and the wood. However, the brick is butted right up to the wood, and water is flowing toward the base and the siding and has caused serious rot. If this corner of the building is opened up, serious damage will most probably be uncovered. The damage must be repaired and the brick removed and reset to provide drainage away form the building. Also, the siding must be raised to prevent contact with the brick.

4. Notice any holes in the siding.
5. Study all window, door, and other penetrations for good, tight joints and caulk.
6. All trim and corner boards must be tight and well caulked.
7. Check siding for rot, termites, beetles, and mold.
8. Check paint and other finishes for integrity. Look for any exterior damage to your framing.

 *Call the Doctor* 

Any large crack, separation, or destruction of siding requires an expert opinion and perhaps opening the building to examine the actual cause of the problem.

Types of Siding

One of the oldest quests for human beings is the search for siding. Many times roofs and siding have both been made of the

FIGURE 6.3

Stucco can be stately and long-lasting, but because it is a solid sheet of a cement mix with wire in it, it must be preserved with care. Separation at door and window frames, watering at the flower beds, and cracks in the stucco itself can allow water to be trapped behind the material and draw termites, mold, and severe rot to the framing.

same materials, but not always—the direct pelting of a roof by the rain requires a more rugged surface for draining water than the siding. Nevertheless, it is very important that siding be durable, skillfully installed, and maintained scrupulously in order to thwart very complex framing repairs. (See Fig. 6.6.)

In the United States, we currently use a number of materials for siding: brick, vinyl, aluminum, stucco (a cement mix that is spread onto wire stretched tight on the wall), and wood products such as boards and shingles. From wood products, which are vulnerable to rot and insects, to materials such as the newer vinyls, which are highly resistant to almost all the elements, the gamut of siding choices grows with every decade.

FIGURE 6.4

Damage begins slowly. Overwatering the bottom of siding can cause the paint to begin to deteriorate, and the longer it is left untended, the worse the damage grows. In this case some scraping, touch-up paint, and adjustment of the watering practices would probably correct the existent damages—but if it continues, mold can grow, the siding will deteriorate, and rot and termites can destroy framing.

FIGURE 6.5

Bushes against home can hold water even if the lot tends to drain away from the building. Note the substantial growth of the vine on the building prohibits siding, frame, and foundation monitoring. The large vine growing up the siding can hold water against it, promoting rot, termites, and damage to framing.

Naturally, there are many variations in the ways in which any of the siding products react with nature. For example, vinyl may be a long-lasting product when it comes to damp rot or insect infestations; however, it can still be vulnerable to wind, sun, fire, etc. The several basic groups of material—earth and stone, forest products, metal sheeting and shingles, and polymers such as vinyl, and even the most durable such as concrete block—all have their strengths and weaknesses. Even a block building needs some attention. Let's take a look at the various products and how to include them in your maintenance plan.

POINTERS

Keep up with siding maintenance on a regular basis. Leaks through the sides of buildings tend to trigger more expensive-to-repair damages than any other type of water intrusion.

POINTERS

Use the checklists to ensure diligent maintenance of wood siding. All wood need some regular attention—especially if water remains in contact with it.

WOOD SIDING

Wood is one of the oldest products used for siding. Along with relative resources such as bushes, grasses, and tropical leaves used for thatch, trees have been used since humans first began to design shelters.

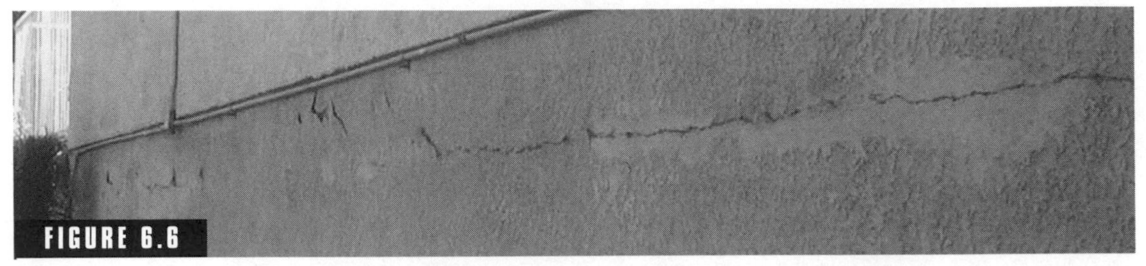

FIGURE 6.6

Structural crack is obvious on siding. Long, deep horizontal cracks that cause destruction of the siding product can indicate severe framing and/or foundation defects.

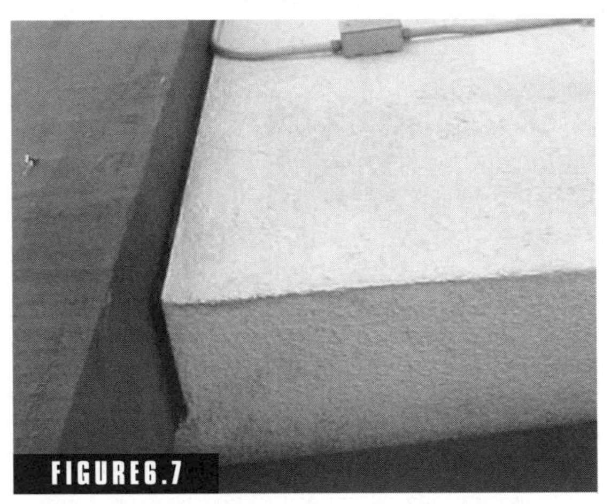

FIGURE 6.7

Beware of all joints in siding. Two different surfaces on a wall or at a corner can trap water seriously. Brick to aluminum, wood to stucco, rock to copper—it makes no difference what type of building you have—the consequences of not keeping joints watertight can be very expensive.

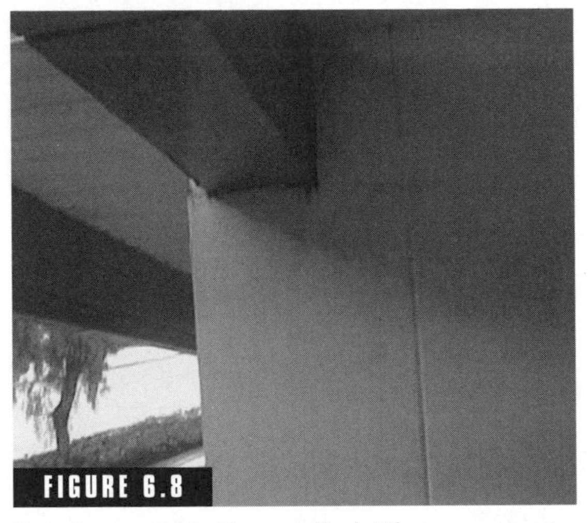

FIGURE 6.8

Keep beam penetrations caulked. Wherever any exterior beams penetrate or attach to a structure, the joint can cause very expensive damage if not well maintained.

Wood products can be an excellent covering for structures such as homes and small commercial buildings. The unique appearance of wood can be a coveted look for design professionals.

The products come in a variety of formats: lap siding, V-rustic siding, board-and-batten, plywood, shingles, pressboards, strand boards, particleboards. Each must be maintained with care. If they are available, follow the manufacturer's specifications.

Wood sidings must be maintained diligently because they can move (swell and shrink) during changes from wet to dry periods. This seasonal motion of the siding units can crack paint and other protection layers; if cracks continue, moisture is trapped under the paint surface. The membrane holds the moisture, causing more rapid decomposition than if the wood were left bare so it could dry out after wet spells.

The Hardiplank drawing is a cutaway view of a lapped-siding installation with a weather barrier. If plain, wooden lapped siding is not cared for, water can be trapped on framing materials and below carpet, causing severe rot, mold, and fungus growth.

Vinyl Siding

Vinyl is one of the many plastic materials that have come into use in the last century. Most vinyl siding is now made so that its elasticity when heated and its susceptibility to sunlight have been taken into account when it is manufactured. Well-made vinyl can be a long-lasting product. It can even be made from recycled plastics. (See Fig. 6.13.)

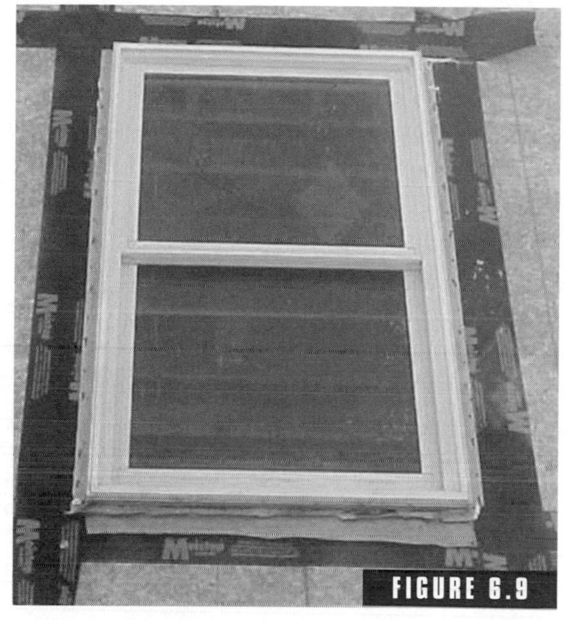

FIGURE 6.9

Moisture barrier at windows is not designed as a true seal to protect the building. Note how the siding and trim will butt up against the window frame. This is where you are looking for a well-caulked joint when you walk your building. Breakthroughs in design have given us polymer (plastic) sheeting for the windows. It appears to be a good deal more rugged than sisal paper—but it is not yet clear that time will not have an effect on the products.

Metal Siding

Numerous metals have been used for siding: lead, copper, stainless steel, etc. Aluminum is very popular with widespread installations. It comes as shingles, lapped siding, and ribbed seams—a host of products that enjoy long life spans.

Most aluminum siding is factory-painted. The paints and the aluminum itself are both durable, but they are also vulnerable to chemicals in the air, which can cause deterioration of the products and movement, which can allow water intrusion into the building. As with all siding products, a regular maintenance program is important to the welfare of the structure. See Fig. 6.14.

Pointers

Even though aluminum has a long life span and does not rust, maintenance is important: Siding, windows, doors, etc., can stay good-looking and avoid deterioration.

Stucco

Among the numerous types of siding (metal, plastic, wood, etc.), stucco is enjoying a real growth in popularity. Originally it was used because wood was not available in the arid parts of the country. But now it has spread to many areas. It comes in a number of styles, such

FIGURE 6.10

Examples of water damage to siding are spalling (chemicals in water affecting concrete, making it disentegrate) and mold infestations from overwatering. No matter what kind of siding, excessive moisture from sprinkler systems is one of the most rampant causes of damage to our buildings. Note the structural cracks and that the earth slopes in, toward the building, where ponding can cause the foundation to settle. In this case, examining the siding or the foundation can signal the owner that there are problems.

as plain, troweled stucco, and even built-up panels, with insulation included as one the of layers of sheet material.

Stucco is a long-lasting product; it is durable and highly resistant to weather; but like all parts of a building, it must be maintained. Stucco is vulnerable to the movement of the building; it cracks and separates from surfaces such as window frames. These situations must be repaired or carefully caulked, preferably during the dry season of the year.

With stucco, follow the maintenance program you put together, and you will go a long way toward preserving your home.

BLOCK, BRICK, AND STONE

Brick and stone can both create excellent, long-lasting structures. Typically they are used as veneers, but they can also be used to build solid walls (Fig. 6.15). Many new synthetic products that resemble stones

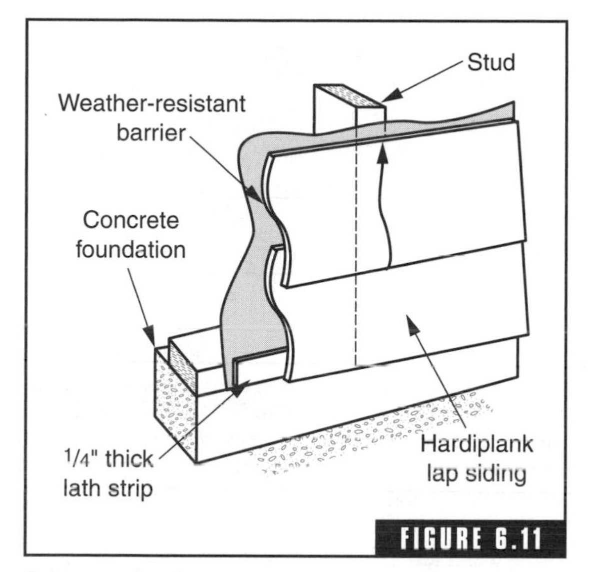

FIGURE 6.11

Cutaway of a fiber cement siding application illustrates the use of Hardiplank®. This is a fiber cement product that is environmentally friendly and long lasting. The drawing is courtesy of James Hardie® Buliding Products. For more information go to http://www.jameshardie.com.

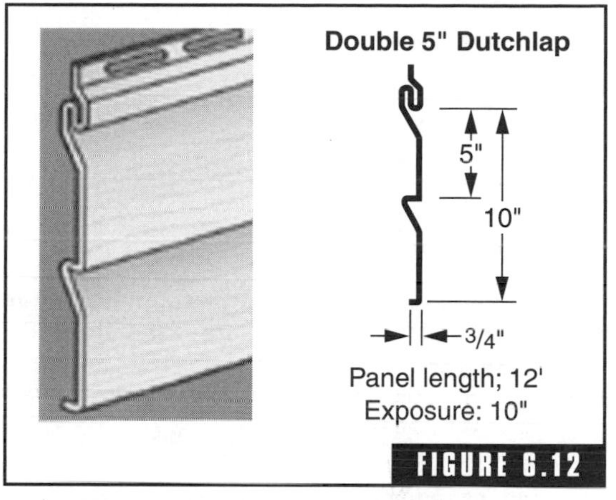

FIGURE 6.12

Dutchlap is one vinyl siding profile that resists rot, termites, and most threats of deterioration. However, all products call for sensible maintenance. Photo used with the permission of Owens Corning. Check out the Owens Corning web site: http://owenscorning.com.

and brick are coming on the market regularly and will probably continue to grow in popularity as the natural products become scarcer and the new generations of designers continue to experiment with synthetic products.

Concrete block is a very handy building product, and it is used on a regular basis. In many areas it is the most versatile and reasonably priced of all products. Block makes a durable skin for a building. Because of its stable mass, it is excellent for maintaining a level temperature inside the building throughout the year, hot or cold.

Naturally, it is highly resistant to the elements as cement and stone can be, but they all require maintenance at the door and window opening to prevent water intrusion and rust, rot, and corrosion at the windows. Also, it is important to tend to any surface moisture.

The Quick Scan

The siding survey is fairly straightforward, although it may require overview from a professional if there are add-ons, changes in materials, or strange penetrations.

POINTERS

The window openings in siding should be checked for secure caulking before every rainy season (Fig. 6.16).

FIGURE 6.13

Today's sidings can include a layer of insulation underneath the covering. Note the insulation below the vinyl siding in the illustration above. The photo is from Progressive Foam Technologies, Inc. (http://www.fullback.com). Check with them for product information. Maintenance is simple: cleaning, caulking any required joints, and making sure that the products are replaced if they are damaged.

BEFORE (Note addition) AFTER EVERBRITE PROCESS

FIGURE 6.14

Aluminum siding will fade and chalk but can be refinished and maintained indefinitely. It will not only look better but it will maintain or increase the value of your home. The picture on the left shows aluminum siding with chalky oxidation and the after picture on the right shows refinished aluminum siding the siding was cleaned and refinished with Everbrite™ Protective Coating which will not only renew the color and finish of the siding but will protect it from the damage from sun, salt air corrosion, acid rain, rust, moisture, and other damaging elements. Homeowners can clean their siding and wipe Everbrite™ on to renew and protect their metal siding, window extrusions, exterior metal furniture, and more. For more information go to http://www.renewsiding.com.

Quick Scan Outline—Siding

Keep these three thoughts in mind while you explore the siding on your building:

1. It is a skin that drains water (this process is absolutely essential—it can save so much in damages that the payment for a well-seasoned professional is inconsequential).
   a. All siding is in place.
   b. Holes in siding are sealed up.
2. Earth movement and buildup.
   a. Earth line is a minimum of 6 inches below the bottom of the siding.
   b. Earth is sloped away from the bottom of the siding.
3. Exterior add-ons.
   a. Check for areas where water may be trapped.

Crawl Space in the Attic

1. Water-stained siding
2. Mold on siding

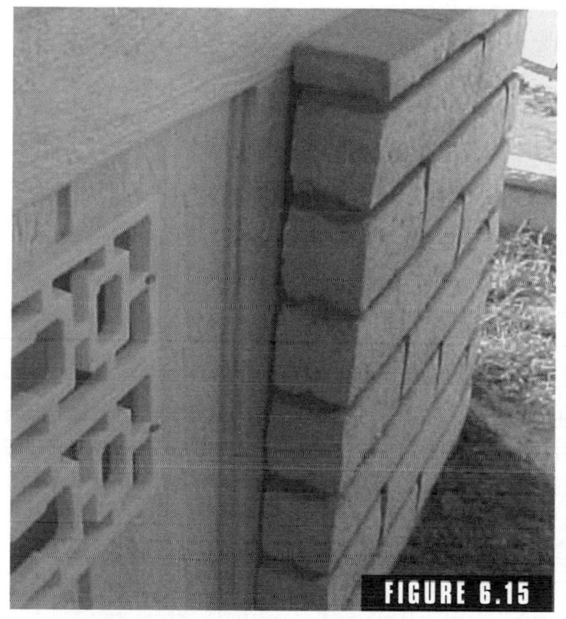

FIGURE 6.15

Joints at stone veneer can trap moisture and rot siding. The framing can also rot, and mold can be generated because the moisture remains for long periods of time without drying out. Note the gap between the factory-manufactured wood and the cement-stone. Any gap in the siding is serious. Also note that even a harmless-seeming opening such as the decorative, white breather to the left of the above photo can trap moisture where the frame of the breather presses against the siding. Keep it caulked, and inspect the area around it during the seasonal inspections of crawl spaces.

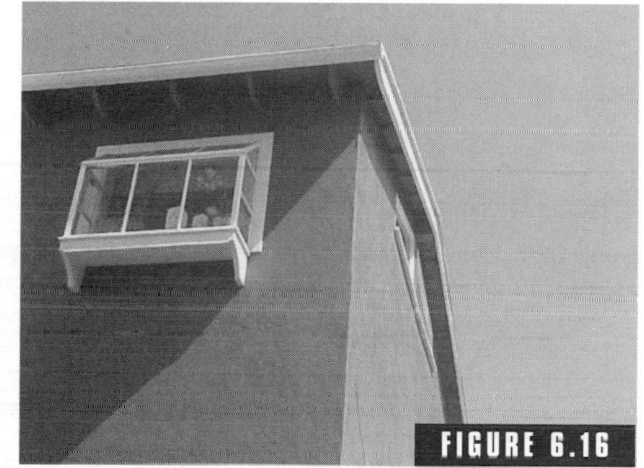

FIGURE 6.16

Special care of hard-to-reach siding penetrations is a must. Note the difficulty of reaching these windows. In a situation like this, the painter will often be the only person who ever examines the joint around the window frame. Most likely, this will only take place every few years. Though the atrium and windows are somewhat protected by the eaves, there will be blowback for sure—and any rot in these areas will be very expensive to repair.

3. Rot on siding
4. Damp siding
5. Holes in siding

Crawl Space or Basement below Ground

1. Water-stained siding
2. Mold on siding
3. Rot on siding
4. Damp siding
5. Holes in siding
6. Siding in contact with earth
7. Siding buried by earth or other materials

> **Pointers**
>
> **Siding damage—especially sprinklers spraying siding on a continual basis—is some of the most expensive to repair that ever happens to a building.**

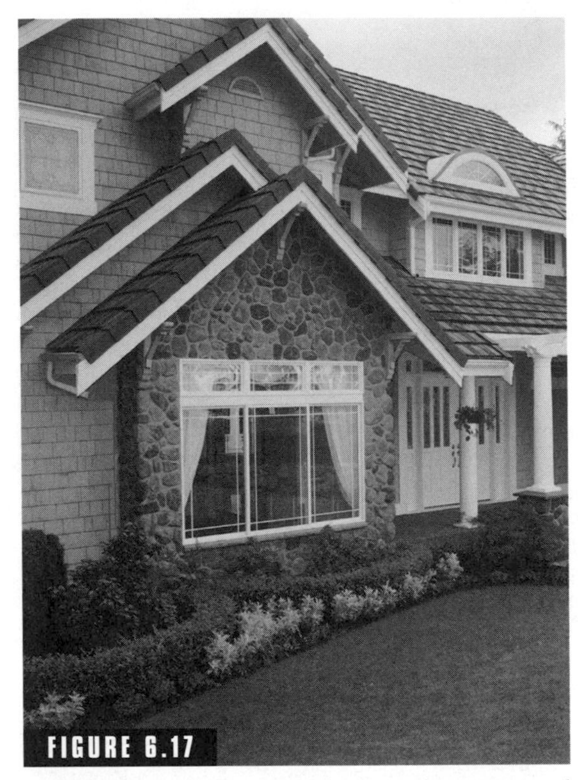

FIGURE 6.17

Manufactured stone veneer is becoming more and more common. It offers the pleasant appearance of stone without a number of installation and upkeep drawbacks. The photo (used with the permission of Owens Corning) is an Owens Corning product from their Cultured Stone® brand product line. Check it out at www.owenscorning.com.

Outside, looking at your home, you will begin to get a sense of the siding as a skin, a membrane that covers the skeleton (the framing) of the building. As you learned in the framing chapter (Chap. 4), it is impossible to understand the extent of problems that may be hidden by the siding. But you can study what needs to be sealed on them to keep mold and rot from spreading.

It is very important that you take this part of your home seriously: door and window openings, foundation, planting at the edge of your home (Fig. 6.19), the roofline at the gutter area, and the siding (Fig. 6.20). If this area of your home is tended to faithfully, the vast majority of repairs can be avoided.

The professionals can discover many problems that you would walk right by. These can be recognized, probed, repairs launched, and tight maintenance plans put in place without allowing future fix-it bills to mushroom into massive debt against the parcel of real estate.

It is pretty simple to remember that what concerns you is water getting in, behind the siding, and being trapped so it never dries out and causes the usual problems—rot, mold, and termites.

At the Earth Level

- Stroll around again, just as you have done before. The more you do this, the better you will get at spotting suspicious situations.
- Earth piled up against the siding at the foundation causes megamillions in damages every year. Make sure the sprinklers are not keeping it wet.

> **Pointers**
>
> **_Remember,_ sprinklers spraying siding cause many hundreds of millions of dollars in damage every year.**

On the Siding

Look over the walls with care. Penetrations, corners, and trim boards—wherever water could get past the skin—are a call for you to check it out. (See Figs. 6.21 and 6.22.)

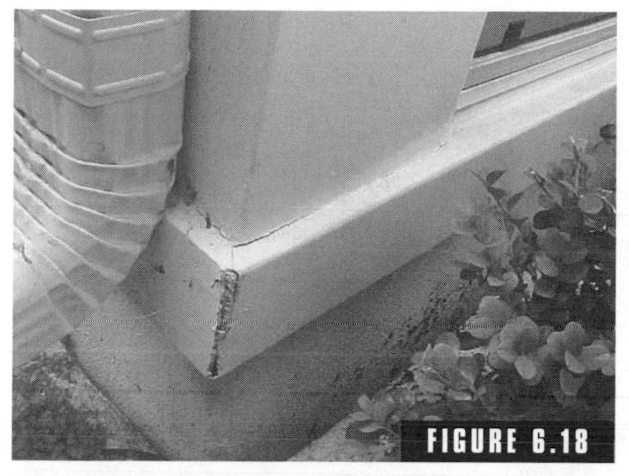
FIGURE 6.18

Very minor problems require attention. The caulk seal adjacent to the gutter downspout is broken. This corner will deteriorate if not taken care of, and the water will infiltrate to the framing. Rot, mold, and termites will follow.

FIGURE 6.19

Vegetation will damage siding by holding water against the frame and building up leaves until they cover the siding. Then the usual problems occur.

At the Door and Window Level

Be aware of any problems where the siding and the door and window frames join. This is very serious.

Higher Up on the Building

Survey all the levels of the structure—the foundation, areas around the door and window lines, the line where the roof joins the siding, the gutters, and the siding up higher, at the attic spaces. Stop and look closely.

On the Inside

Just as with the framing, when you are back inside, go down (or send your experts) into the crawl spaces, and slither up into the attic areas, looking for signs of damages, dampness, and all the usual.

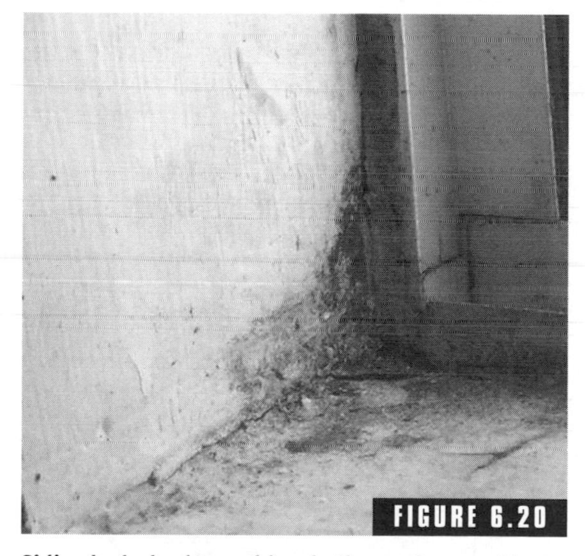
FIGURE 6.20

Siding leaks lead to mold and other serious problems. Keep an eye out at door and window frames—sometimes siding leaks are first seen as damage inside the building.

Developing the Siding Checklist

Try to think about the big picture when you look over your siding. There are two main concerns related to the skin that have to be

FIGURE 6.21

Watch for peculiar junctions of different building materials—protecting them from water intrusion can be a big project. Notice the stucco is cracked, the wood is separated from it, and all the siding dies into the brick and concrete, which means that they are suffering from water intrusion. Planning the fix for this type of situation requires the input of an expert.

addressed year after year: (1) The covering of your building must stay solid and must not be left to decompose, and (2) all penetrations through the siding on buildings must be caulked when any rain, snow, or sprinklers are watering the structure. The problems related to the siding are very expensive to repair: massive rot, intense mold infestations, and structural shifting.

POINTERS

Siding with rot or termites requires immediate repair.

 *Call the Doctor* 

If you are in doubt about anything that penetrates the siding of your buildings, get an expert. Expensive repairs can be triggered.

The Siding Checklist

Done	Item	Call a pro	Notes
✓	**SIDING**	✓	
	TYPES OF SIDING		
	Solid wooden v-rustic		
	Solid wooden lapped		
	Solid wooden board and batten		
	Solid wooden other		
	Plywood		
	Manufactured wooden		
	Wood shingle		
	Aluminum		
	Vinyl		
	Other		
	All doors and windows caulked tightly at joint between door frame and siding—all the way around the opening		
	All penetrations in siding sealed tightly		
	All paint and protective surfaces for siding in excellent condition		
	All flashings painted		
	All rot contained		
	Sprinklers not watering siding		
	Debris clear from siding		
	Earth 6 inches below siding(minimum)		
	Water draining away from siding		
	Roof not draining on siding		
	Other		

FIGURE 6.22

Post will trap water—watch for all objects on the siding that will hold water behind them.

FIGURE 6.23

Watch all joints with mixed siding.

If you are in doubt about anything on any of the checklists, remember that *every* line item is a crucial part of the home, where expensive repairs can be triggered.

Bringing It All Back Home

As we have suggested, the siding material is literally the skin of a building. Whether it is wooden, synthetic, stucco, brick, stone, or a combination of materials, if it is well designed and applied, it keeps the elements away from the building's interior components and the inhabitants. However, as mentioned above, even the most long-lasting and well-applied siding requires ongoing maintenance.

One of the most vulnerable parts of the skin of a building is the holes in it, which are required for doors, windows, and ventilation (Fig. 6.24). All the openings into a built structure are vulnerable to the elements on a continual basis. They need cleaning and caulking regularly. Draining systems can clog and hold water within the structure. The siding materials can crack, move, warp, and rot, allowing water onto and beneath the building's water barriers. Then the frame and interior finishes of the structure can be exposed to moisture.

A building's skin needs to be surveyed frequently, caulked, repainted, and maintained. The siding section checklist is for monitoring the various places where the sides of the building are in need of care. With the lists you continue to build your maintenance plan and establish a pattern for finding obvious problems such as missing or old caulk as well as other situations that require your attention.

The primary concern when you are creating a practical siding maintenance program is to shift into the habit of thinking of the siding as a continuous membrane that protects the building—after all, it just looks like a bunch of wood or bricks or stucco, not anything like a living skin. As with most membranes, there are openings in the siding, and they must be built correctly and then watched with care on a regular basis to keep the siding from deteriorating and allowing the elements into the structure, which will cause it to deteriorate.

FIGURE 6.24

Beware of new penetrations—any penetrations in siding must be flashed and sealed, or else rot, mold, and termites will become a serious problem.

The siding is often painted. It is very important to keep paint solid as a membrane—if water is getting inside the membrane, the paint has the opposite effect of its original purpose—it seals water in against the frame and sheathing materials, not allowing it to evaporate in a timely fashion. This promotes rot, rust, and the other deterioration problems that a building is susceptible to, as well as mold and fungus growth.

The main things that we are looking for when we set up the siding checklist are breaks and potential breaks in paint and siding products. They typically occur at the various joints around the building, such as the place where the window frames join the siding material at the window openings.

You will walk the building, taking a look at the entire surface all the way up to the roofline, and evaluate what type of openings, penetrations, and other places for water to enter are present. As you continue to work on maintenance, you may become more and more sensitive to anything that causes damage to your buildings. You will automatically watch for buildup that can hold moisture against the base of structures, which causes rot, water intrusion, mold and fungus growth, and infestations of destructive pests such as termites (Fig. 6.25).

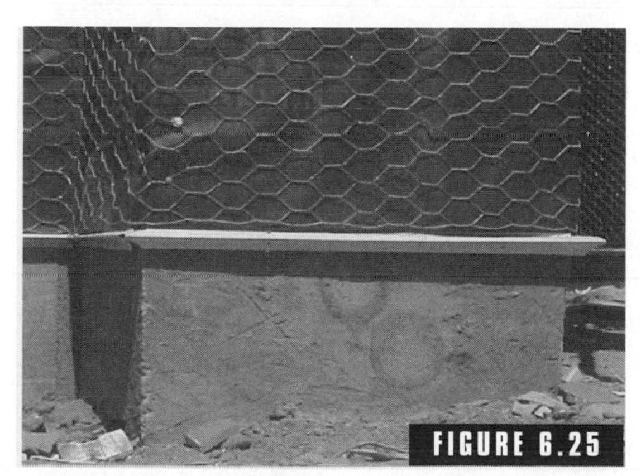

FIGURE 6.25

Keep siding well above the earth. Even if things start off great as in this picture, the buildup can be dramatic.

POINTERS

Keep a close eye out for buildup of soil and plants next to the perimeter of your buildings at ground level. This condition can trap water against the siding and is often the cause of a great deal of unseen damage—a 6-inch space is code, 12-inches is much better.

The Siding Resource Directory

Aluminum and vinyl siding on historic buildings. The appropriateness of substitute materials for resurfacing historic wood-frame buildings; vinyl siding on historic buildings.
http://www.oldhouseweb.net/stories/Detailed/228.shtml

Boulder Creek Stone. Manufacturer of stone veneer, thin brick, floor tile pavers, and accessories. Stone veneer, thin brick, and floor tile pavers.
http://www.bouldercreekstone.com/

CDI—artificial stucco repair. See how to repair artificial stucco so you can get a repair bond from your termite company.
http://www.cdisite.com/reviews/stucco/index.htm

Cleaning copper. What solution can be used to clean copper siding from brown oxidized to shiny quickly (i.e., with one or two wipes)? Also how can the red color that results from overexposure to certain acids be developed?
http://www.finishing.com/133/98.html

Concrete block homes. In Florida's hot, humid, hurricane-prone and insect-laden climate, homes built with concrete block make perfect sense. However, northern builders often dismiss this medium, saying it limits design flexibility. Not true, say the Florida builders.
http://www.growinglifestyle.com/article/s0/a66162.html

Concrete-foam block houses resist 150 mi/h winds. Appeared March 26, 2000.
http://www.dominionpost.com/a/realest/2000/03/26/000326d/

Cultured Stone. Owens Corning Cultured Stone, an Owens Corning company, is the world's leader in producing realistic artificial stone products.
http://www.owenscorning.com/around/exteriors_new/stone.asp

Design review guidelines—corner commercial buildings. Exterior siding. The retention of the frame weatherboard and/or wood shingles is essential for historic houses in the East Row Historic District. Modern materials such as aluminum or vinyl siding, imitation stone, or imitation brick are not permitted.
http://www.eastrow.org/design_review/siding.html

El Rey, manufacturer and distributor of stucco wall systems. Manufacturers of stucco and EIFS wall systems in the southwestern United States.
http://www.elrey.com/

Evaluative summary of articles on randomized block design. The authors wanted to test how people react to new product advertising. They examined in a 2 × 2 × 3 randomized block design whether experience or format affected the response.
http://www.indiana.edu/~jopeng/Y603/art5B.html

Everbrite Protective Coating. It prevents tarnish and corrosion on brass, copper, silver, steel, etc! Save hours of polishing to keep your brass and copper looking great.
http://www.everbrite.net/TarnishPrevention.htm

Hummel Industries, Inc. Specializing in cut stone and the restoration of masonry and concrete, Hummel Industries, Inc., has been a leader in the industry since 1847.
http://www.hummelindustries.com

Irish Stone Walls and Stone Buildings—Contents. Narrow your search: arts, crafts, mosaics, stone business, industries, construction and maintenance, materials and supplies, masonry and stone, cast stone business, industries, construction and maintenance, materials.
http://homepage.tinet.ie/~mcafee/

La Habra Stucco. Major supplier of stucco and related products.
http://www.lahabrastucco.com/

National One Coat Stucco Association. Supports the manufacturing and use of one-coat stucco. Learn what the material is made of and why it enables design flexibility.
http://www.nocsa.org/

Preservation Brief 8: Aluminum and Vinyl Siding on Historic Buildings. Preservation briefs assist owners and developers of historic buildings in recognizing and resolving common preservation and repair problems prior to work.
http://www2.cr.nps.gov/tps/briefs/brief08.htm

Revere Copper Products, Inc. Producer of roofing copper: traditional flat sheets and a variety of coil products, including specially shaped coils for labor-saving pan formers and steamers, gutter and leader coil, and soft roll copper for flashing applications.
http://www.reverecopper.com/

Sakala Stone Products—handcrafted stone for your home. Handcrafted stone can be applied to any structurally sound surface made of wood, wallboard, block, brick, concrete, or metal.
http://www.sakalastoneproducts.com/frames.htm

Schundler Product Guide—Perlite In Simulated Stone, Masonry, and Wood Products. Basic guide describing how perlite can be used to

produce simulated stone, masonry, and wood products.
http://www.schundler.com/sim-stone.htm

Siding. Home Tours Links Events History Business Guestbook Guidelines Membership Village Voice Guidelines—Siding. As some of the area's oldest structures, German Village's frame buildings accent its historic character.
http://www.germanvillage.com/GVguide/design/existing.html/

Siding systems. Tolley Hughes Inc. official Web promotion site. All about what we do in roofing sheet metal and architectural design. Maintenance and free estimates. Custom design and job-specific assistance available.
http://www.tolley-hughes.com/metalwalls.html (MSN)

Sierra Nevada White Granite. Looking for the finest-quality natural granite stone at guaranteed lowest price? Located near Reno, Nevada, White Granite is producing beautiful landscaping and masonry stone at below-wholesale prices.
http://WhiteGranite.com

Stuccolustro.com. Stucco majestro is a stucco lustro technique, which was already used centuries ago for the coating of inner surfaces.
http://www.stuccolustro.com/

Stucco Pro. Offers moisture testing and inspection nationwide. Order the company's video, entitled "Stucco: Real Problems, Real Solutions."
http://www.stucco-pro.com/

Stucco siding tips. Learn about basic features, common problems, and easy repairs.
http://homeadvisor.msn.com/improve/siding/stucco/intro.aspx

Synthetic siding. A building's historic character is a combination of its design, age, setting, and materials. The exterior walls of a building, because they are so visible, play a very important role in defining its historic appearance.
http://www.staunton.va.us/cityhall/histdist/hdres1j.htm

Synthetic stucco litigation. Information page for the Synthetic Stucco (EIFS) Class Action and Senergy and Thoro Settlement (Ruff et al. v. Parex, et al. No. 96-CVS-0059). If you own or formerly owned a residence with synthetic stucco exterior wall cladding, your rights may be affected by the national settlement of a class action lawsuit.
http://www.kinsella.com/eifs

Texas Stone Designs, Inc. artificial stone veneers simulate limestone and cobblestone as well as hearth. A quality manufacturer of artificial stone veneers which have the look and feel of real stone including marble, flagstone, cobblestone, and hearthstones for fireplaces. http://www.texasstonedesigns.com/

Union Station Brick. Since January 1985, Union Station Brick has gained a reputation for success as Nashville's leading distributor of brick, mortar, and sand and superior building products.
http://www.home buildersdirectory.com/union%20station/default

Vinyl siding maintenance and energy and historic preservation. Wood must be painted or stained, but vinyl siding needs only a yearly washing to maintain its fresh appearance. Keep in mind, however, that window sashes and wooden trim will still require routine painting.
http://architecture.about.com/library/weekly/aavinylb.htm

CHAPTER 7

Roofs

Because of their out-of-sight, out-of-mind location, roofs are one of the most neglected parts of our built structures. Although they are not entirely enclosed, as is the framing, roof care very often drifts into the background of the building owner's concerns (Fig. 7.1).

Combined with their out-of-the-way location, they are dangerous to inspect and repair because they are high off the ground. And as if that isn't enough, it is very easy to damage roofs seriously by activities on the roof deck. They can also be damaged by neglect of tasks as simple as not cleaning gutters or walking on the roof deck. Unless you are a very learned, dedicated, and determined handyperson—or a roofer—it is imperative that you use a roofing professional for both inspections and repairs.

To the novice, roofs are very complicated because they are difficult to access (Fig. 7.2) and understanding the interweaving of all the shingles and felts and flashings is a bit much. To help you get a handle on what makes your roof safe, there are a few major concepts to think about.

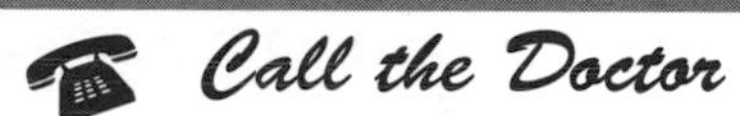

Never put off roof repairs. Always make sure that you trust your roofer and get all repairs done immediately. If there is doubt about your expert, bring in a third party right away.

Most all roofs have penetrations that allow light and air into the building and vent gases out of the structures. These openings are extremely vulnerable to the weather all the time. If not maintained, they allow water into the building. Along with planned penetrations, problems such as

FIGURE 7.1

Roofs are very vulnerable. Simple neglect, such as not cleaning gutters, can lead to many thousands of dollars of unnecessary repairs.

broken and missing shingles and punctures in the roofing materials cause a lot of damage, fast (Fig. 7.3).

Roofs need regular inspections. The speed with which damage occurs is remarkable because roofs are so directly related to the elements. They require caulk, repainting of metal flashings and other roof-deck products, cleaning, and replacement at regular intervals. This chapter explains those conditions. With the checklists for different roofs it also presents suggestions for the owner to engage the use of a roofer to inspect the building twice a year.

Jump-Starting

Roof maintenance is absolutely critical, but it is also straightforward because unless you

FIGURE 7.2

Danger on roofs is found everywhere: loose tile slips, felt paper can rip, debris can settle on flashing, tiles break, shingles tear, and falls can be fatal. Roofs are the domain of skilled professionals. Note that the galvanized flashing is not painted and that it will rust completely through and cause extensive damage if it is not maintained.

are very knowledgeable and handy, you will be using a roofer to get you started and check things out in preparation for every wet season.

There are just a few straightforward tasks involved with keeping your roofs tight and ensuring their long lifetime. (See Figs. 7.4 and 7.5.)

1. Lock in the service of a very good roofer who has spent a lot of time working on your type of roof. This means that you do not want a tar-and-gravel specialist working on your slate roof. Get a true slate craftsman.
2. Keep it inspected and clean, including the gutters. And keep a good coat of high-quality, exterior metal paint such as Rustoleum on flashings that aren't extremely long-lasting like copper.
3. Repair your roof promptly.
4. Replace roofs regularly.

Keep those ideas in mind as you think about your roof. And remember, as always, to keep it simple. With these ideas in mind let's move on to the plan.

FIGURE 7.3

If you see anything suspicious during your site walks, show it to the roofer and get it fixed immediately. Look at the photo above closely. Note that the primary rot is in the lower corner of the roof. There doesn't appear to be such intensive destruction up higher, above the roof rafter (larger 2 × 4 board that is on edge, holding up the roof). This could mean that debris from the trees sat on the roof in this area for long periods of time and/or the gutter was not cleaned in this area and the cellulose material from the trees above held water against the underside of the roof for long periods of time.

The Plan for Roof Maintenance

1. Read this chapter and study your roof—identify what type it is.
2. From the ground, fill out your checklist, but stay off the roof deck.

Use a roofer. Do not inspect your roof. Do not repair your roof. Stay off your roof and keep others off unless your roofer is watching them. Only the very skilled handyperson should attempt her or his own roof maintenance—it is dangerous, and just walking on a roof can easily cause expensive-to-repair damage.

FIGURE 7.4

Maintenance can absolutely save money. Note the replaced board in the photo. This is a minor repair that illustrates how easily rot can spread from something as harmless seeming as not cleaning your gutters. If the roof is not maintained and the gutters are not kept clean, the rot and swarms of termites can spread over the entire roof and into the frame of the building.

FIGURE 7.5

The importance of roof maintenance cannot be overstated. Not only can rot be an intense problem in the frame of your buildings just below the roof, but also the water can course down the walls, drenching other areas of the building, and can stay in place for long periods of time without drying up.

POINTERS

Set up a close arrangement with a roofer for yearly inspections. Late summer is the most important time: the year's wear and tear on the roof has taken place, and the roofers aren't extremely busy, so they can be more relaxed and look at your home closely. Plan to pay the roofer for this work—you always get what you give. And remember, this meeting does not take care of cleaning your gutters after the leaves fall in the autumn.

3. Use word of mouth to find a reliable roofer who specializes in your type of roof.
4. Review your checklist with the expert.
5. Schedule an inspection with the roofer for the beginning of each wet season—this includes snow as well as rain.
6. Edit your checklist, let the roofer review it, and file a finished copy in case there is a need to change roofing experts.

Types of Roofs

Pitched Roofs

The pitch of a roof is the angle at which it rises from the flat surface at the top of the wall where the framing stops. In other words, if you stand in the street and look at the line where the siding stops and the roof begins—usually there are gutters at this point—you will see a rise from the framing line to the top of the roof (see Fig. 7.10 for a photograph of a pitched roof).

The main design concept behind pitched roofing is that the steeper the angle, the faster the water flows off the building, thus adding life to the roofing materials, the roof framing, and the rest of the building. However, if a roof has a radical pitch to it, installation and maintenance become very slow and difficult—notice that most buildings near you have a moderate pitch.

There are several departures from the ordinary trussed roof with a ridge at the top and a slope down to the gutter line on either side. Shed roofs are the main variation used in ordinary residential construction. Naturally the name arises from the shed shape—one wall is higher than the other and when the roof rafters are placed, they take on the rise from one wall to the other. The main advantage to the shed roof is that it is very easy, and quick for the carpenters and roofer, thus saving noticeably on labor.

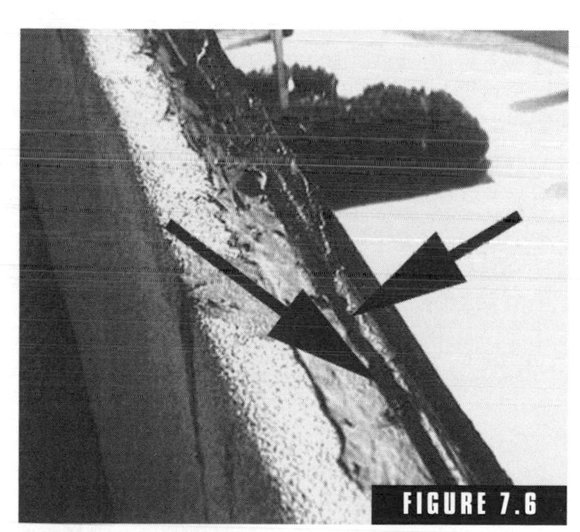

FIGURE 7.6

If there is any doubt about your roof, get more opinions. The out-of-sight, out-of-mind syndrome makes roofs very vulnerable, so you have to be able to trust your roofer. Note the suspicious buildup of tar and gravel and the joint (the arrows are pointing at the joint between the asphalt, roll roofing, and siding) on the right edge of the roof. Does it look suspicious? *It is.* A take-charge, highly knowledgeable roofing professional is required.

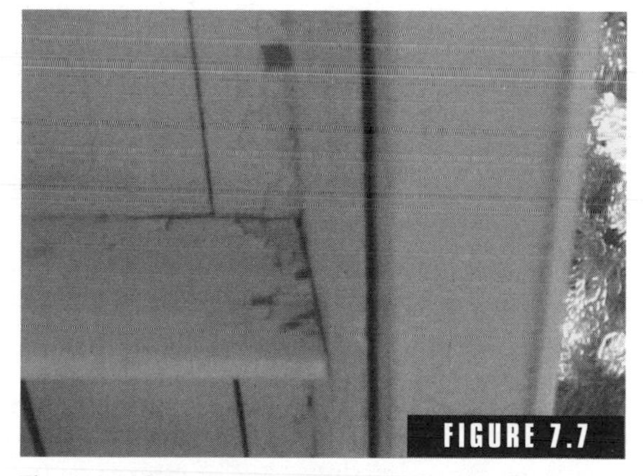

FIGURE 7.7

The ground tour can expose problems. Show your roofer everything that you are suspicious of while you do the "Roofing Checklist" together. If it all looks too confusing or you catch yourself dragging your feet, get the roofer over to help you put the checklist together. Asking the roofer questions will help you ascertain your comfort level with him or her. If you don't feel good, trust your intuition—those initial reactions are often the clearest.

Bad gutter joints leak behind siding. Note that at the spaces between tiles and over the lip of the tiles, water can work its way down behind the gutter. Gutters are very important on your home. They are not a place to scrimp. Use your roofer if the skills are there—put in wide, deep, and ample gutters. Walk the property yourself and examine them. If anything looks doubtful, get a second opinion.

FLAT ROOFS

Flat roofs also have some rise from the plane created by the top of the wall framing and a fall to the roof drains. This slant is typically gradual, but it is very important that water is not left standing on built-up roofing because of the wear and tear on the surface and the fact that if there are any punctures (which frequently take place), water will leak into the building.

The design community tends to shy away from the residential flat roof. They do not add the interesting spatial variation for design elements that are offered by pitched roofs. For commercial buildings, they are the mainstay because they are materials- and labor-efficient, as is the shed roof.

Types of Roofing Products

ASPHALT SHINGLES

These are the mainstay of home roofing in the United States. They originated as a composite of fabric, fillers, and minerals used to help

POINTERS

Some roofing products that are overly weathered on the top, such as delaminated slates, can absorb water on the underside as well as the weather side, hold it, and cause rot to the frame below. Make sure the expert checks for damp top surface to roofing felt contact.

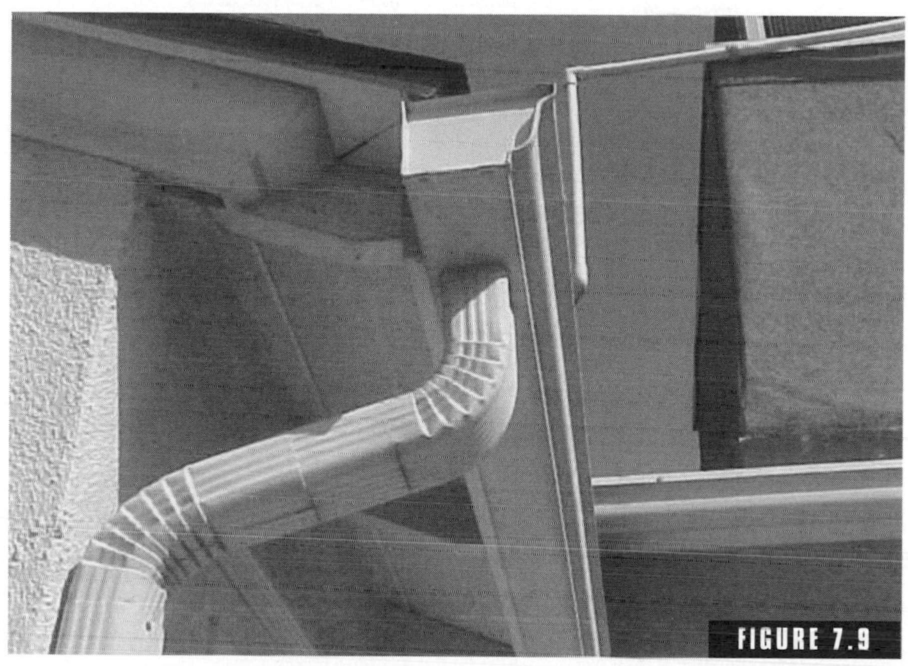

FIGURE 7.9

Well-tended gutters look solid clean and tight. However, looks can be deceiving. This can put you in the out-of-sight, out-of-mind mode, and there can be buildup in a gutter. Notice how the roof sheathing (wood on the roof that the shingles are nailed onto) rests right on the top of the gutter—water can easily soak the lumber, and then rot sets in quickly.

shield the petroleum materials from the sun and the elements. More recently, manufacturers began to produce shingles using a fiberglass mat to replace the felt. Because of its rugged nature, fiberglass tends to have greater tear and fire resistance and to be easier to handle than earlier products.

During your walk around the building, the main things to look for on asphalt

If you are a person who must inspect and work on your roof yourself, it is highly advised that you use a roofing professional to coach, at least for the first project that you do on the roof deck.

POINTERS

Never put off a roof inspection. **It can save you thousands of dollars in home repairs.**

POINTERS

Never inspect a clay tile roof unless you have years of experience. **Even the most handy person needs to learn to navigate clay tiles—they are VERY easy to break.**

shingle roofs are buckling, separations, tears, thin gravel, damage from branches and leaves fallen from trees and people on the roof deck, and any other irregularities that appear to be of concern (Figs. 7.10 and 7.11). Your roofer will help you with the rest.

CLAY AND CEMENT TILE ROOFS

Clay tiles are still similar to the original, old mineral product that our ancestors dug from the earth near their homes. Clay tiles are more uniform, but they are still extremely vulnerable to people walking on the roof deck. In spite of the ease with which clay tiles crack, they can provide a wonderful, long-lasting roof of great beauty. See Fig. 7.13.

ROLL ROOFING

Roll roofing is an old product that works just fine. It can provide a roof that drains well and is easily maintained. One of the finest things about roll roofs is that they are easily replaced—the new product can go right over the old with little preparation. This reroofing can continue as long as the roof framing can support the weight of the layers.

The main things to look for are buckling, separations, tears, thin gravel, tree droppings, and any irregularities that appear to be of concern.

FIGURE 7.10

Asphalt shingles can make a stately roof. ***(Photo used with the permission of Owens Corning.)***

SLATE ROOFS

Because of its long lifetime, some may think that slate is invincible. But maintenance is important with slate roofs just as with any others. Like clay tile, it is one of our earliest roofing sources. Found all round the world, it is a large initial investment but with a life span of up to 200 years, it can actually be a cost-effective solution. See Fig. 7.13.

Delamination is one of slate's drawbacks, on both the upside with the weather and the underside. Reacting with the air, the slates endure a molecular change into gypsum and disentegrate very slowly.

Wood Shingles and Shakes

Another early roofing material, called *shakes and shingles,* was a very handy way to protect the home when we lived in the proximities of unlimited forests. Today, they are more limited in their use due to the spread of suburbia and the universal availability of artificial products such as asphalt shingles.

Wood shingles and the more dense shakes can produce a very stately roof, and a great many strides have been made in the direction of fire safety. If you have a wood roof, it is important to keep a close eye on it for splitting, curling, and rot.

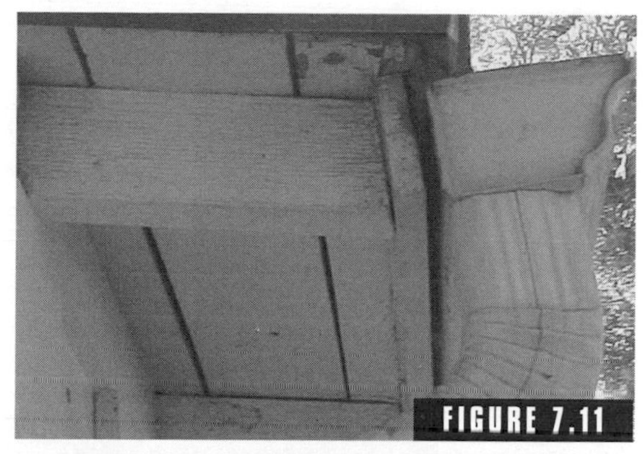

FIGURE 7.11

The wood and surrounding areas look clean and solid on well-tended gutters (see Fig. 7.9). But gutters that have not been cared for tend to be surrounded by wood that is discolored and looks as if it is suffering from rot or other problems.

The Quick Scan

The first thing to do is to take a look at what you will pay attention to when you are out looking at the roof.

Quick Scan Outline—Roofs

You will be exploring three main concepts while looking at your roof:

1. Shedding water
   a. Visible roofing is solid, no missing parts.
   b. Roof is clean so water can flow off.
   c. Flashing appears to be in place.
2. Complete gutter system
   a. There are no missing gutters.
   b. Gutters are clean.
   c. Downspout drains water away from foundation.
3. Roofer found

If You Have a Crawl Space in the Attic

1. Water stains on the roof frame
2. Termite infestations

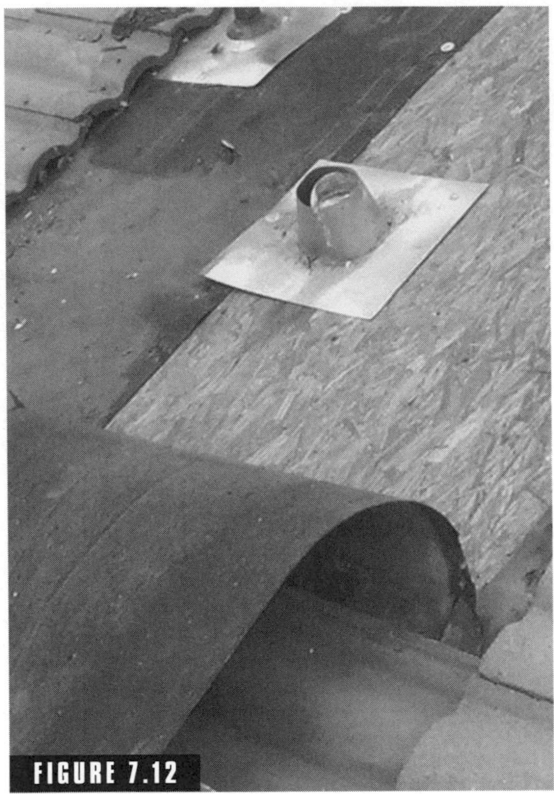
FIGURE 7.12

Many penetrations in roofs must be maintained to avoid serious leaks into your home. Note the thick, rugged quality of cement tiles. They are much less vulnerable to breaking than clay tile, but amateurs still need to stay off the roof deck because they are prone to damaging them. Note the aluminum flashings at the penetrations through the roof. These will not be visible unless fixes are required on your roof, but a really good roofer will spot problems and suspect defects that are covered, for example, damaged flashing, mislapped felts, and loose tile.

POINTERS

Most experts recommend that slate roofs not be replaced until 15 to 25 percent of the roof needs repair. This is a rule of thumb, and decisions should be made with a true slate expert.

3. Mold growth
4. Signs of rot
5. Damp spots
6. Separated framing joints
7. Water stains on the roof sheathing (Fig. 7.14)
8. Light through roof.

Now let's get out for a good look at your roof. Go across the street and walk up the block. Don't rush yourself—check it all out in detail. Notice if it is flat or pitched (built at an angle) and what products it is covered with: wood shingles or shakes, cement tiles (flat or Spanish style), slate, clay (Spanish) tiles, or asphalt tiles.

Walk back down the block in the other direction, then through the side yard and into the back. After you have looked at how the whole top of the structure works from a distance, walk the building from a closer vantage point. Be sure to see if you can spot any visible damage. See Figs. 7.15 to 7.17.

As you tour the front of your building up close, study the covering itself, looking for problems with the roofing system: cracked tiles, slipping shingles, missing flashing, debris on the roof deck, anything suspicious to point out to your roofer (see Fig. 7.26). Check out the gutter system closely, and see if you can tell if it is in disrepair, if it is complete, and where it drains. See Figs. 7.18 and 7.19.

Now continue to walk the side of the structure and into the back yard. Move as far away from your home as you can get, and take a close survey. Then move in closer and repeat the same steps you took in the front yard. Don't worry about repetition. All these different appraisals will give you an in-depth understanding of where troubles arise. Not only will you begin to be able to spot areas that need some care, but also you will get better and better at being able to understand what professionals are doing and saying, and with attention you will learn how to make certain that they are doing their jobs.

FIGURE 7.13

Slate can last so long that it is actually very economical. Steep pitched roofs tend to last longer than lower sloped roofs. This photograph is courtesy Mr. Jim Germond from Brandon, Vermont. He can be reached online at the Liza Myers Gallery: http://www.lizamyers.com for more information. We found the photo on the New England Slate Company web site http://www.nesslate; check it out—carrying on the old American traditions.

FIGURE 7.14

Leak stain from the many roof penetrations. When you inspect the roof area, notice water stains and rot—they are often visible. Bring your professional's attention to them during the inspection.

FIGURE 7.15

Skylights are vulnerable to water intrusion—they can be an area that is rampant with serious rot, mold and termites. Their maintenance is a must when you are setting up your plan.

FIGURE 7.16

Rust has started on flashing—the arrows to the right are pointing at deep rust that will spread and eat right through the flashing quickly. Soon it will become a penetration, and the water will destroy the roofing felts and start to work on the frame of the building.

FIGURE 7.17

Flat roof and deck drains are very vulnerable: Water can back up, rust through, and disintegrate flashings; and all the dangers are very close together and affected by each other. Rust, rot, mold, even termites can all be readily activated if flat roofs are not tended regularly and with professional overview.

FIGURE 7.18

Gutter dumping on foundation. Note that as soil builds up and shifts, the immense amounts of water flowing from this downspout can cause tremendous damage to the frame of the building, mold infestation, as well as possible foundation settlement. Make sure that your gutters drain well away from the building.

FIGURE 7.19

Odd roof structures must be inspected by professionals. This is very important: Condensation from air conditioners, movement of the structure, and failure of flashing—many things can damage the building.

 **Call the Doctor** 

Even if you don't do all the work to maintain your home, get your roofer over each year in the late summer.

Developing the Roofs Checklist

When you are thinking about the maintenance of roofs, the main question is, What happens after water hits the building? From the top, the water must flow over the shingles or tiles, or down the slight drainage slope of a tar and gravel roof. The roof must allow it to flow smoothly and quickly and be able to shed all the water that hits it into the gutters and downspouts. From there, it is very important that all the water drain off the lot, or into well-designed, on-site drain cisterns. Keep this in mind while we work on the checklist.

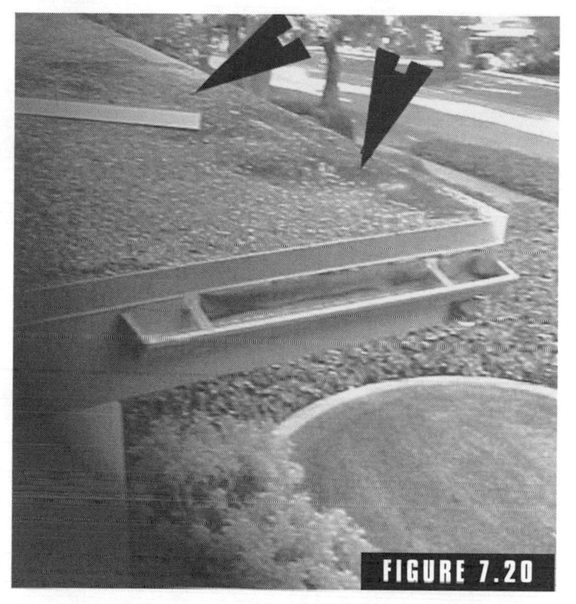

FIGURE 7.20

Over time built-up roofing can loosen the gravel that protects the tars from weather (see dark spots at arrows). It can dam up gutters and allow too much heat gain from the sun. A professional roofer can spot defects quickly. It is best that you stay off your tar and gravel roof unless you have a thorough understanding of your type of roof or are very handy and want to do a great deal of research and take responsibility for damages you may cause by walking on the roof deck.

The Roofs Checklist

Done	Item	Call a pro	Notes
✓	ROOFS	✓	
	TYPE OF ROOF		
	Asphalt shingle		
	Cement tile		
	Clay tile		
	Shake		
	Wood shingle		
	Slate		
	Mansard		

Done	Item	Call a pro	Notes
	Flat		
	Mixed		
	Other		
	GENERAL INFORMATION		
	Age and condition of the roof		
	Damaged and missing slates		
	Should be repaired		
	Should be replaced		
	Delamination		
	Slate tiles holding water		
	Other		
	FROM THE GROUND		
	Debris		
	Clean gutters		
	Clean valleys		
	Solid ridges		
	Cracked, broken, missing tiles		
	Rusted, unpainted flashing		
	Environmental pollutant stains on roofing		
	Algae or growth on roofing		
	FROM THE ATTIC		
	Sweating		
	Ridges		
	Valleys		
	Flashings		
	Gutter line		
	Other areas		
	Water stains		
	Mold and fungus		

Done	Item	Call a pro	Notes
	Dampness		
	Rot		
	Daylight		
	Cracked tiles		
	Sufficient attic ventilation		
	Rot		
	Other		
	GUTTER SYSTEM		
	Downspouts		
	Other drains		
	Daylights on foundation		
	Drained to street		
	In good repair		
	Other		
	AFTER VISIT WITH ROOFER		
	Inspection plan set with roofer		
	Special notes from meeting		
	Other		

Bringing It All Back Home

No matter what kind, with a little care your roof can last quite some time. Even roll roofing on a shed can last for years, and with several roofs placed right over the original, it can serve you for decades.

All the roofs must get close scrutiny every year. Putting this off leads to problems such as flashing that has pinholes, rust, and built-up dams

> **Call the Doctor**
>
> **If you decide to do anything related to your roof that requires a ladder, be very careful. Ladders can be extremely dangerous in the hands of amateurs. Make certain that it is level and secure both top and bottom. Inspect each part of it for good repair, and set the angle so your work position is safe and comfortable. If you don't know what is safe and comfortable, get help. It is preferable to have a person on the ground to tend your ladder and help by handing you anything that is required from the ground or roof deck.**

FIGURE 7.21

Slate can last well over a century under the correct conditions, but maintenance is important. If you have slate, make sure the expert you choose has years of experience with the material.

of debris and is no longer watertight. Your roofer has a far better chance of finding damaged roof products or those that are missing or loose or out of sequence than you do alone. No matter if it is slate that is delaminating, cracks in tiles, or bad drains on tar and gravel, seasoned, competent professionals will often see the problems quickly.

Trying to inspect a roof yourself is very iffy. It takes years to notice things quickly. When the problem is compounded by the damage you will probably do by being on the roof deck and the dangerous situations that walking a roof provides, you will probably be best served by hiring a pro.

Be sure to flag your roofer to watch for damage or blocked up gutters and downspouts. If the roofer doesn't do your gutters, bring the gutter specialist into your maintenance plan.

In the attic, wood rafters and sheathing should be checked for water stains and rot at the typical areas such as valley joints and the intersections of roof planes to vertical walls. Your expert should be well aware of these types of leaks simply from fixing lots of them.

Your maintenance plan will include regular cleaning of all gutter systems and repairing and replacing anything quarrelsome that is spotted on the roof deck.

Be sure to keep a clear log of all expenses for roof inspection, maintenance, and repair as well as any there work. Just like the upkeep records for your car, this can be a valuable sales tool when you sell the building.

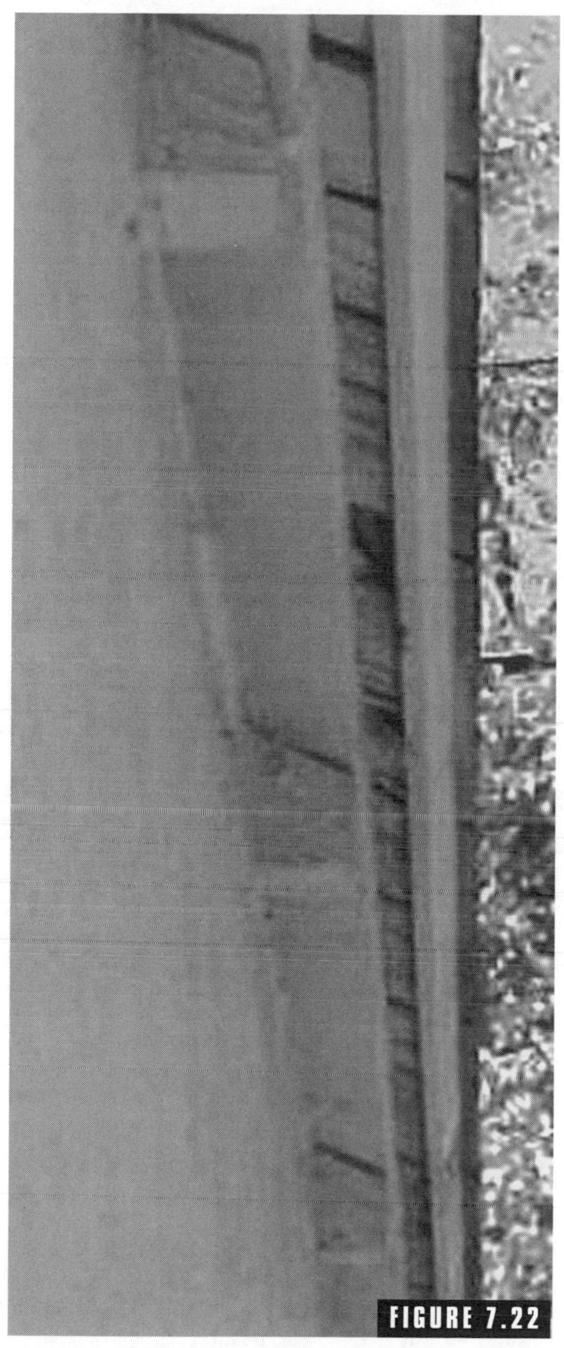

FIGURE 7.22

After several layers, roofs require a complete rebuild. Look closely at the roofline on your building; the built-up layers of roofing products are often visible. Be sure to ask the roofer if there is any doubt.

Asphalt Shingles

The majority of readers will have asphalt shingles, which are the most abundant residential roof (Fig. 7.23). They are comparatively lightweight and easy to apply, and they have a reasonable life span. There is a wide selection of colors readily available, and they are easy to repair.

The roofs are very simple to maintain, and they drain well. Another major asset is that they can be roofed over easily—the new shingles go right over the old roof without intense labor. Simple roof replacement works fine as long as the framing supports the load and water damage is not a problem.

However, it is important that your roofer be highly experienced and know when to strip the roof, replace any damaged framing, and start with brand new shingles.

Clay, Cement Tiles, Wood Shingles, Shakes, Slate, and Other Roofs

Stay off the roof deck—only a seasoned professional knows where it can endure the weight of a person. Talk with your specialists about

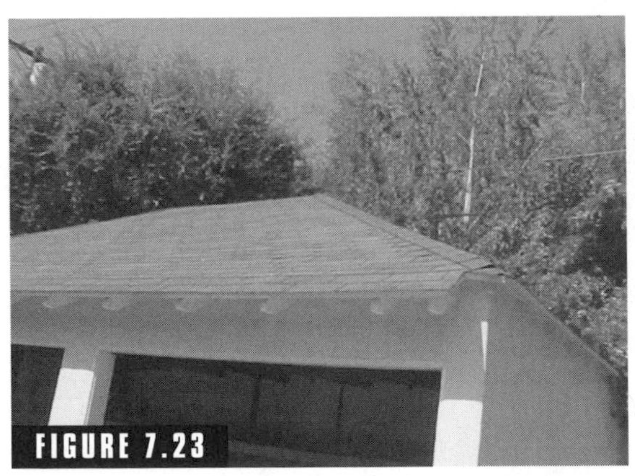

FIGURE 7.23

Old roofs with asphalt shingles can be very reliable. If they are well tended, repaired when needed, and replaced at the appropriate time, asphalt roofs can stay tight for decades on end. Note that the front of this garage does not have a gutter; that can work fine if it drains onto a driveway that takes the water right out to the street. However, the side is not guttered either. Again, if there is parking and water is not leaking between the foundation and the parking slab, this can be a fine way to drain water away from a building. But if water is pouring off the building toward the foundation, serious problems can occur.

FIGURE 7.24

Debris (at end of arrows) in valleys. This is the flashing that drains water from the areas where different roof planes meet on sloped roofs—they are called *valleys*. Debris can gather in valleys, creating dams that cause water to flow back under the roofing felt, the flashing, and the shingles. This leads to all the usual problems including rust cutting right through the flashing. Note that the flashing is galvanized and can definitely rust—it should be painted and kept painted. On very expensive roofs such as slate, it is wise to pay for copper flashing, which will last the life of the roof.

making building paths or platforms for people who must be on the deck. This is beyond your capabilities unless your are extremely well versed in the craft, or intend to put in many hours of study and possibly spend a lot of money on trial and error.

POINTERS

Do not plan to flip slate tiles for reuse. Delamination happens on the underside of roof slates.

You must be very careful when putting your own or any other weight on these products. Inspect them from the ground or a ladder, using binoculars. For workers who have to use the roof for their tasks, set up a scaffold or platform and walkways to spread the weight.

Cracked, broken, misaligned, and missing shingles, and the degree to which delamination has occurred on products such as slate, should be noted, as well as damaged flashings and broken or clogged downspouts. Photographs can help in discussing your concerns with contractors. In

FIGURE 7.25

When roof framing dies at a wall, the joint is vulnerable. Note that this area of any home is prone to water intrusion into the building. Flashing will be placed on the roof and up the wall. Roofing paper is interwoven. The siding and roofing products are then applied, and all this must be done with accuracy. Part of good maintenance then demands keeping the flashing clean so dams are not created, backing up water under the roofing and soaking the siding.

FIGURE 7.26

This drawing is from GAF Materials Corporation's Web site. For some great pointers about your roof, go to *http://www.gaf.com* on the net. Click on Residential Homeowner, then Learn About Roofing. You will see the drawing above and more information. Tour the whole site while working on your roof checklist.

the attic, note water stains and rot on wood rafters and sheathing. Problem areas are usually near the roof framing junctions and at the intersection of roof planes, such as at valleys and hips (Figs. 7.24 and 7.26).

Regular maintenance should include cleaning gutters in the fall while leaves are falling if there is rain—a good cleaning after the leaves have fallen—and in the early spring before the rains begin, and replacing damaged areas promptly.

The Roofs Resource Directory

Asphalt Shingles

Hammer Zone—replacing individual asphalt shingles. Learn the techniques of replacing damaged or weathered asphalt shingles, given detailed instructions and photographic illustrations.
http://www.hammerzone.com/archives/roof/repairs/replshingle/

Insulation, roofing, siding, acoustics, and composites. Owens Corning offers insulation, roofing, siding, basement and acoustic systems for building and remodeling, composites solutions, and asphalt.
http://www.owenscorning.com/

Cement Tiles

Columbia Roof Tile. Concrete roof tile made from sand, cement, and pigment.
http://www.columbiarooftile.com/html.html

Elagante Roofing Tiles. This company makes unique cement tile in both slate and shake styles. For sloped roof, steep roof, tile, roof tile, cement tile, flat tile, hard tile, hard roof tile, hard roofing tile.
http://www.elagante.com

Galloway Roof Tile. Concrete interlocking tiles used for pitched surfaces are available in 10 smooth-finish colors. The range consists of three tiles and the Galloway, a thin-leading-edge flat tile that looks like a slate when laid.
http://www.russell-rooftiles.co.uk/crossproductpages/rooftiles-galloway.asp

Miracote, fracture-resistant underlayment for use under ceramic tile, quarry tile, marble. Waterproof deck coating.
http://www.miracote.com/mirfrre5.htm

Roofing: Ludowici roof tile, standard fittings, homeowner. Roof with a tailored, polished appearance provides significantly better protection against leakage than mitered pieces held in place by cement.
http://www.marshallwindows.com/croof/crlu00901p.html

Roof Tile Institute. Cement, pigments, sealers, and adhesives necessary for tile manufacturing and installation.
http://www.ntrma.com/factsht.htm

Roof Tile Management, Inc. Fiber cement slate roofing products page (aka fiber cement slate). Fiber cement slate, slate roofing, roofing slate, roof slate, fiber cement slate. Samples of projects undertaken by Roof Tile Management using fiber cement slate roofing products.
http://www.rooftilemanagement.com/Products/Roof-Tile-products-slate-fibre.htm

Tile roof material. Installation Guide of roof tile industry. Installation of concrete and clay roof tile for moderate-climate regions of North America.
http://www.wkspunk.com/ceramic-kitchen-tile.htm

Clay Tile

Preservation Brief 30: The Preservation and Repair of Historic Clay Tile Roofs. Qualities of a clay tile roof. Inspection of the roof structure and the roof covering is recommended. Repair of the tile roof.
http://www2.cr.nps.gov/tps/briefs/brief30.htm

Roofing Tile and Its Application, Part 6: Lining the Roof and Applying Tile. The Tile Man presents an educational series of articles about the design, manufacture, and installation of clay tiles.
http://www.thetileman.com/art6.html

Roll Roofing (Asphalt)

Atlas Roofing—roll roofing products. Roll roofing materials.
http://www.atlasroofing.com

Lo-Slope Roll Roofing. Easy-to-apply roll roofing membrane is specially designed. Grand Manor Shangle Celotex premium laminated shakes metal roofing systems. Lo-Slope Roll Roofing System.
Info@nationalcontractors.net

Roll roofing products at Tarco. Felt underlayments, including perforated products. Fiberglass mineral surfaced roll roofing is a general-purpose product.
http://www.tarcoroofing.com

Roofs—General Information

Copper gutters listings and resources at the Builders Data Building Products and Services Directory. Welcome to the Builders Data Building Products and Services Network, world's largest copper gutters directory.
http://www.buildersdata.com/diy/gutters/coppergutters.html

Roof Seek.com. This is a consumer roofing resource with roofing-specific links, forums, and auctions; manufacturers.
http://www.martinfireproofing.com

Slate

American Slate Company. Learn about the many interior and exterior applications of slate, and find out how to contact the company.
http://www.americanslate.com/

Evergreen Slate Company, LLC. This company has provided the roofing slate for many landmark buildings. Visit the Famous Roofs page.
http://www.evergreenslate.com/

Preservation Brief 29: The Repair, Replacement, and Maintenance of Historic Slate Roofs. Installed properly, slate roofs require relatively little maintenance and will last 60 to 125 years.
http://www2.cr.nps.gov/tps/briefs/brief29.htm

Slate Roofs in America: A Short History by Jeffrey S. Levine. Philadelphia Historic Preservation Corporation—roofs in America.
http://www.sacredplaces.org/PSP-InfoClearingHouse/articles/

Stone Locator. Limestone, marble, granite, slate, and ceramic tile search engine.
http://www.stonelocator.com

Tile and stone care. Sealers and cleaners for marble, limestone, granite, slate, and grout.
http://www.stonetechpro.com

Traditional Roofing Newsletter. Dedicated to preserving the knowledge and skills of the traditional roofing trades. Issue 2, Number 1, Spring 2002, "Graduated Slate Roofs: Traditional Roofs from Historic Roots," by Joseph Jenkins. http://www.traditionalroofing.com/TR2-graduated.html

Vermont roofing slate. There is a 100-year guarantee for this Vermont roofing slate. Email for samples. http://camaraslate.com/roofingslate

Snow Guards

See http://www.slateandcopper.com and Fig. 7.27.

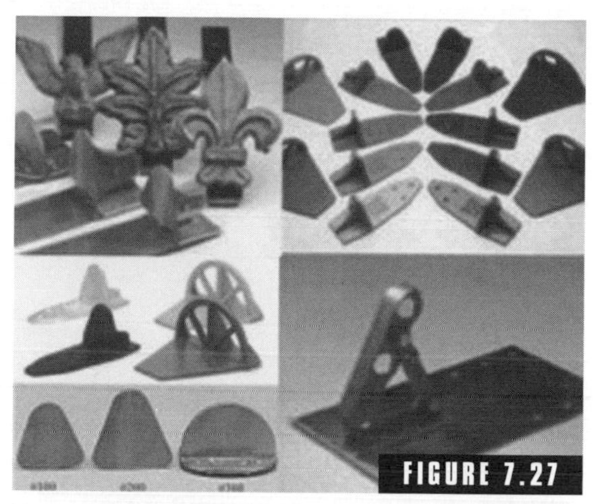

FIGURE 7.27

M. J. Mullane manufactures snow guards and snow management systems for every type of roof surface. They even have retrofit snow guards. Go to the Web site to see all of the Bronze Guard snow guard collection. http://www.bronzeguard.com/

CHAPTER 8

Electric, Plumbing, and Air Conditioning

Keeping an eye on the mechanical parts of your home is fairly straightforward. Getting a handle on the various systems and syntax may be the hardest part for you. Construction professionals will often use different terminology for various parts of buildings and probably won't include all the wire and piping systems in the structure under the heading *Mechanical*—but for this book we want to keep it simple.

The electric system does not tend to damage your home or suffer damages on a day-to-day basis, although it can. Rusted breaker boxes and damages to electrical components related to termites and framing rot can occur. (See Fig. 8.1.)

Water can seep out of the air conditioning, condensation can create moisture damage, and of course the plumbing can trigger major damage to any building. So it is very important that all these systems become a part of the maintenance plan, even though they may not require a great deal of preventive work.

We are not going to give much thought to the minor, offshoot systems such as the security devices because they don't tend to trigger construction damages that keep multiplying into major repair bills.

Our exploration is not about replacing toilets, but is about spotting leaks that may be damaging your home. As the framing is like the skeleton of your body, and the siding is like the skin, so the mechanical systems are like the bloodstream and central nervous system of your home. With few exceptions, the wiring comes in from the street

FIGURE 8.1

Possible danger at an exterior light. Have the electrical expert look for any dangerous electric devices outside as well as inside, and learn to spot them yourself at any time of the year.

FIGURE 8.2

Neglect of maintenance can even cause danger from rust at gas entrance. If there is any sign of danger anywhere along your gas system, including the slightest odor of gas, *do not wait for your expert—call the gas company immediately*. The company will dispatch a highly skilled professional who will be equipped and trained to find a gas leak and render it harmless.

 Call the Doctor 

If there is ever a suspicion of a gas or water leak or danger from electrical, heating, or air conditioning—never wait for your expert. Get your family to a safe place, and call the fire department immediately. Next call the gas, water, or electric company and request that a skilled technician be sent out quickly and at no charge.

to a meter on your site. From the meter it typically flows through the circuit breakers, which are a group of hefty switches designed to recognize a problem and shut off, and out to the rooms.

Plumbing provides the path for a similar routine with the gas and water systems in your buildings. In the majority of homes, gas comes in from the street to another meter and is then dispersed to ranges, water heaters, and other appliances (Fig. 8.2).

Heating, ventilating, and air conditioning (HVAC) dispersal takes place in a similar manner: Air is taken into chilling and purifying devices, conditioned, then dispersed through large pipes to the various areas being supplied.

Water is distributed in much the same way, coming in from the street, flowing through a meter, and moving right on out to the various fixtures in the building. The major difference with water is that it requires a waste system to dispose of it after its use. The grounding devices on the electrical system and the vents for by-products of gas combustion are similar to the waste system for water, but they are not nearly as complex.

Gas and electric are very important because they are capable of inflicting mor-

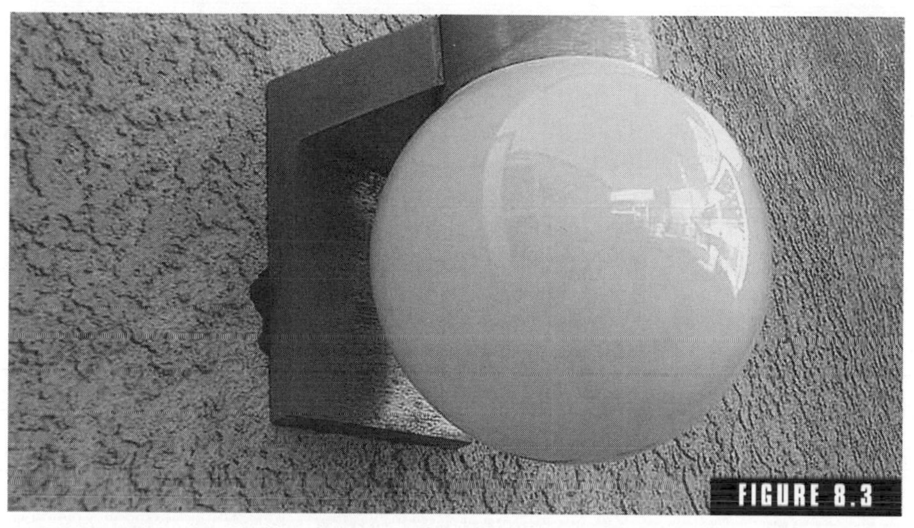

FIGURE 8.3

Hole at exterior light—take a close look at all exterior electrical work. Remember what the interior of walls looks like, what a torrent of water rolling down through the hole in this picture can cause inside your home—all the typical problems that you have studied.

FIGURE 8.4

Side of lamp is properly caulked, but bottom was overlooked. Note that the tight bead of silicon caulk does not proceed under the base of the light fixture; nor does it proceed across the top. It is spread generously over the joint, but the person was not thorough. Consistency is one of the foremost requirements for quality maintenance. This is the type of error that one can find with hired maintenance companies—there is no way to know whether they are doing a good job, unless you monitor the work.

tal damage to inhabitants of a structure. Make sure your experts investigate the systems for danger while you are putting this list together.

Gas and electric can also destroy the entire building readily with fire and explosions, but that is not the real subject of the material in this book. Although condensation from HVAC units can cause damage, the main network that can launch serious, accumulative destruction to your buildings is the water system. The main piping network in your home is continually charged with water, and large amounts flow through it on a daily basis. Obviously, even a small hole in either the piping or the wastewater flowing out can keep the framing wet, basements and crawl spaces surcharged with moisture, and foundations swampy. (See Fig. 8.5.)

Although there are a good many things going on with the mechanical systems, they are not as complex as they may seem. If it all sounds a bit much at this time, just go on through and walk your home again, and work out the checklists. Remember, the more you look closely at your home, the more information will sink in, and in time you can know what is supposed to be going on in every little nook and cranny. Also keep in mind that working through the preliminary checklists, and then the final maintenance plan, with professionals who really

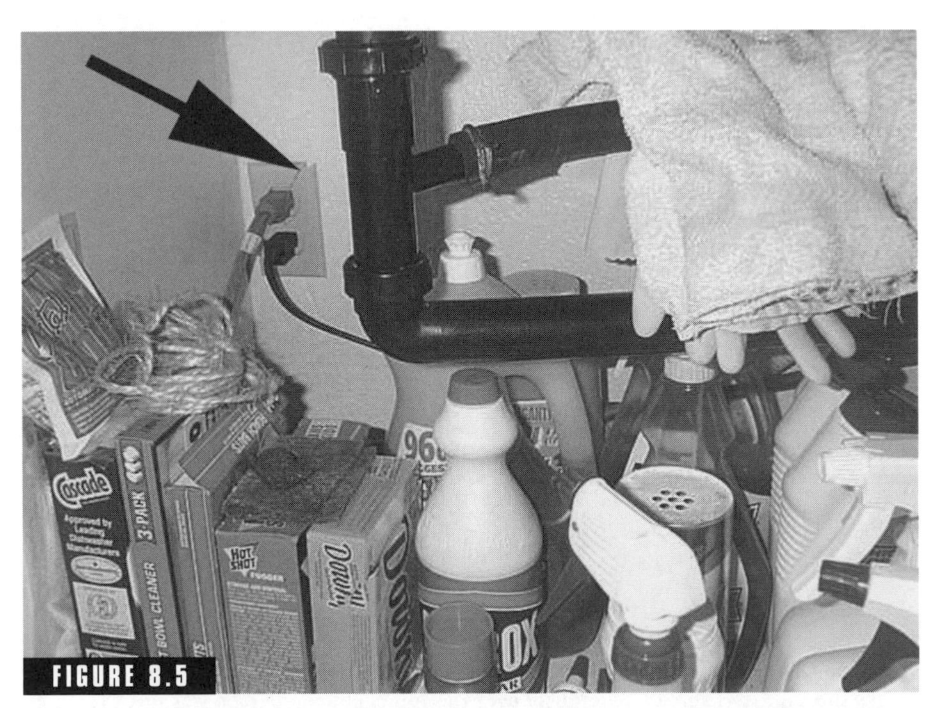

FIGURE 8.5

Broken plate under sink. Have your experts take a look at everything when you set up your lists. Get things fixed and keep them fixed. But the main situation in this picture that could cause long-term water damage would be if the black waste pipes were leaking. This will rot your cabinets and can cause mold and even termites.

understand the way a building works will bring you up to speed much faster than going it alone and catch needed repairs.

Jump-Starting

The Electrical, Plumbing, and Air Conditioning Checklist requires attention to all the systems as you walk the building. Let's take a look at what we are trying to achieve. There are three areas for upkeep of the mechanical system:

- *Water damage.* Make sure that water is not damaging your property.
- *System damage.* Check out your electric, plumbing, and air conditioning sytems for damage.
- *Safety survey.* Check out all the systems for safety, and maintain them.

On the plan below, the basic concepts for putting together the Electrical, Plumbing, and Air Conditioning Checklist are broken out. After you take a look at the plan, we will explore the property for all the components.

 **Call the Doctor** 

Carbon monoxide **is a threat to the inhabitants of a home. Detectors and regular maintenance of all the systems greatly reduce your vulnerablility to this widely present toxin.**

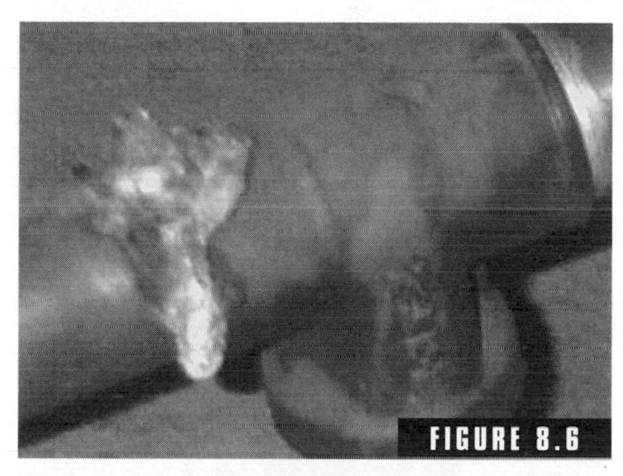

FIGURE 8.6

Buildup on copper pipes. This condition may exist inside the pipes. If you notice and/or have slow pipes, be sure to point it out to the professional who helps you. The residue can slow down, and even cut off an entire water supply.

The Plan for Electrical, Plumbing, and Air Conditioning Maintenance

Exterior

1. Review the "Quick Scan" section.
2. Walk your lot for the electric, water, and gas entrances.
3. Look for any conditions that appear to be dangerous.
4. Look for any indication of standing water from plumbing.
5. Check your HVAC units for rust and any standing water that is breeding organisms.
6. Note anything that you want to ask your expert.

FIGURE 8.7

Penetrations behind tub-shower. Note that if the fixtures on the other side of the tile, inside the shower, are not caulked, water will be running down the inside of the sheetrock, ponding on the floor and causing rot and mold growth. This photo is inside the wall, opposite where you would be standing (right through the center hole in the sheetrock) behind the copper intersection. Inside your bathroom, this condition would be indicated by mold at the side of the tub, spongy sheetrock, or signs of mold and rot on the floor below, viewed from down in the crawl space.

Interior

In the House

1. Look for electrical danger or problems—broken outlets or switches, etc.
2. Check for air conditioning problems—it will probably require an expert.
3. Study all areas where plumbing leaks may occur: under sinks, behind toilets, at tubs and showers.

Attic

1. Look for loose wires, open boxes, etc.
2. Look for signs of mold or condensation stains on and near HVAC ducts.
3. Look for plumbing leaks and punctures in the roof that are leaking.

Basement and/or Crawl Space

1. Scan for loose wires and dangerous junctions.
2. Survey for condensation on HVAC pipes and any standing water at the system that is breeding mold.
3. Study the basement's earth, floors, and walls for wet spots and standing moisture that may indicate plumbing leaks.

The Quick Scan

First, outside on the lot, let's look for the various systems. The water meter is often found out along the curb, near the street. It is usually a rectangular concrete lid with rounded corners. It is often flush with the earth and usually has a small hole for lifting. Call the water company or your plumber for a look at the small, round meter.

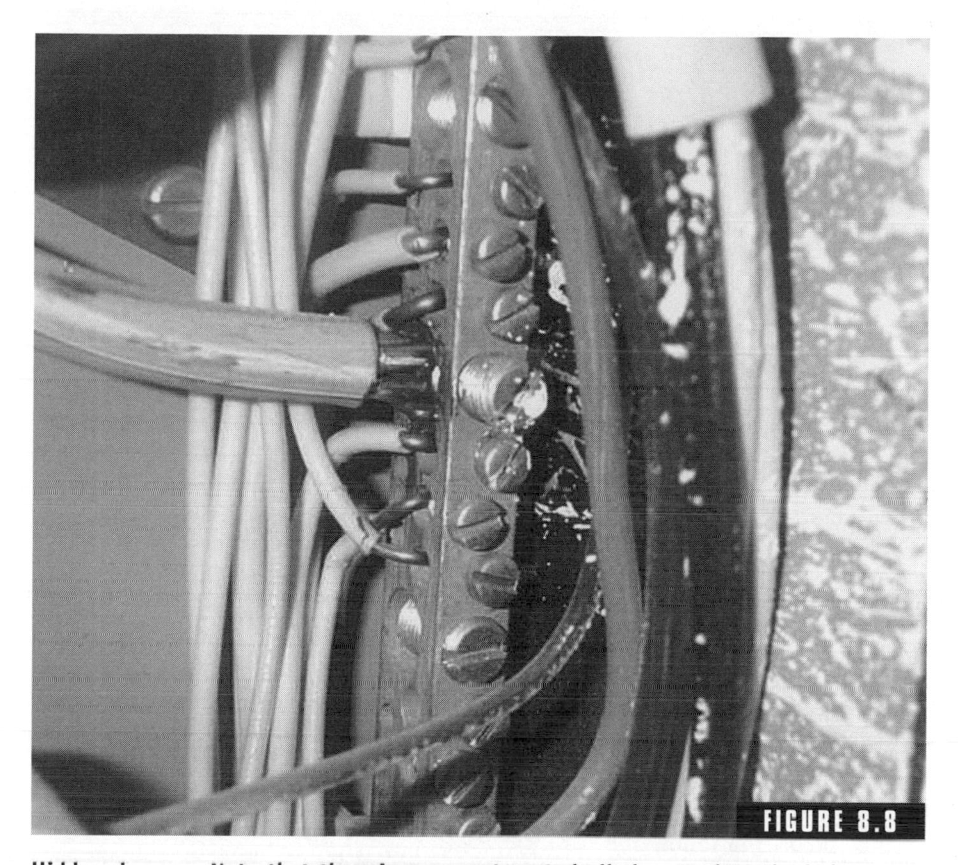

FIGURE 8.8

Hidden dangers. Note that the wires are not seated all the way into the holes up to where the plastic sheathing has been stripped off them. This condition can be dangerous. Because this area of your electrical system is inside the breaker housing, it would require an electrician to examine your box. ***Never unscrew the protective covers off breaker boxes yourself—get the power company or a certified electrician.***

The electrical system typically starts out on a power line from the street, and then there is a service drop to your building. Often there is a mast with a weather head, and the wires drop down inside a pipe to a gray metal box, often mounted on the side of one of your buildings, which houses the electrical meter. Electric meters are about 6 inches across—you will see a spinning disk and a few small meter gauges above. In newer subdivisions, equipment is often underground (see Fig. 8.14)—call the electric company for a tour of your system.

Gas comes in underground from the street in a smallish pipe to the gas meter (see Fig. 8.2). The meter is sometimes found in shed-type

FIGURE 8.9

Wires and fire blocking (the board with the holes in it that deters stiff wind currents inside walls from spreading up through the space between studs to ignite second-floor and roof framing) can be damaged by the electricians who build a home.

POINTERS

If you are not extremely handy or learning to be so, there are several systems in this chapter that require a professional to inspect the property closely, fix problems, and set up maintenance plans. The following are the key places where the experts must be engaged:

- ***Electric.*** **Check the wiring for safety, fix all dangers, set up the electric maintenance portion of your checklist system.**
- ***Water. Leaks*** **from water supply and waste lines, water heaters, rusted water pipes, corroded or leaking copper, interior plumbing leaks, septic systems, repairs, and regular service by plumber.**
- ***Gas. Service*** **entrance area, piping, appliances, repairs and regular service.**
- ***HVAC.*** **Units, filtration, piping condensation and toxins breeding, general service, repairs, regular service.**
- ***Low-voltage systems.*** **Regular inspection and service.**

closets attached to the house, behind doors under the house, and sometimes right out in the open. It is often a gray metal device something like a solid figure eight, with the white meters in the upper area.

Also part of the plumbing system, many water heaters can be found in the home, the garage, and on the exterior of buildings. They are sometimes found in exterior closets, which are in various states of repair.

After you check out the following parts of your mechanical systems, take a look at what you have figured out on your own; then call in an expert. Find the heating, air conditioning, and any air filtration systems. If you have trouble understanding which is which, you need to call your expert from the "Who Does What Checklist" right away.

Exterior

1. Review the "Quick Scan" section.
2. Walk your lot for the electric, water, and gas entrances.

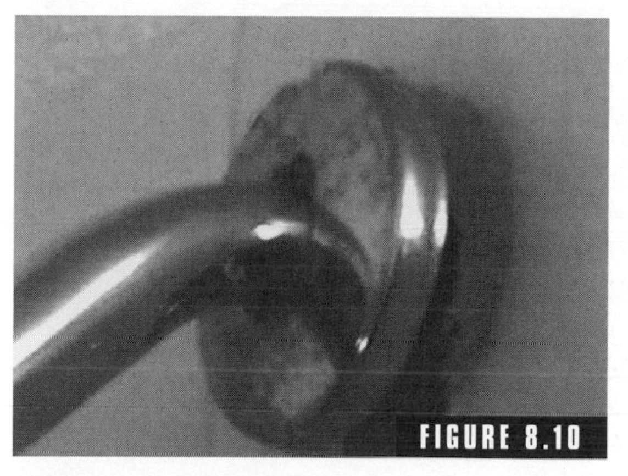

FIGURE 8.10

Missing caulk at penetrations can cause serious damage inside walls.

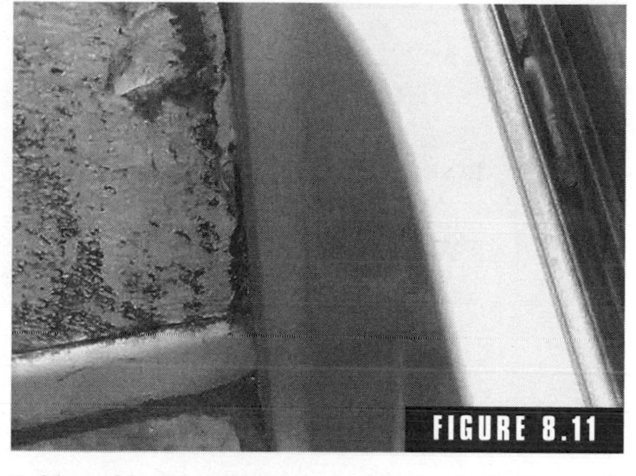

FIGURE 8.11

Mold outside tub and damage to finishes indicate a serious leak, probably a plumbing problem—get help now.

3. Look for any conditions that appear to be dangerous.
4. Look for any indication of standing water from plumbing.
5. Check your HVAC units for rust and any standing water that is breeding organisms.
6. Note anything that you want to ask your expert.

Call the Doctor 

Do not touch metal. **It conducts electricity; if you happen to contact metal while also touching a live wire, current can flow through your body, increasing the chance of dangerous shocks.**

Pointers

When the electrical professional is at your home, be sure to request a close survey of the grounding system for all your electrical work. Make sure that they also inspect for ground fault circuit interrupters (GFICs) wherever there is water and electrical appliances or plugs. Ask them to point out any other danger zones.

Interior

In the House

1. Look for electrical danger or problems—broken outlets or switches, etc.
2. Check for air conditioning problems—it will probably require an expert.
3. Study all areas where plumbing leaks may be: under sinks, behind toilets, at tubs and showers.

Attic

1. Look for loose wires, open boxes, etc.
2. Look for signs of mold or condensation stains on and near HVAC ducts.

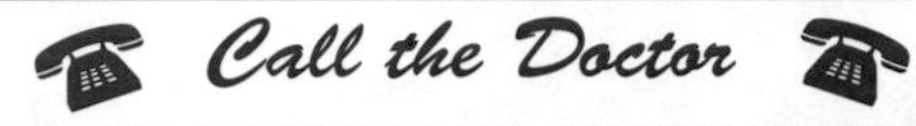

DO NOT ATTEMPT TO FIX CIRCUIT BREAKER PROBLEMS—THIS IS VERY DANGEROUS WORK. DO NOT TAKE THE FRONT PLATE OFF THE BOX. CALL AN EXPERIENCED ELECTRICIAN OR YOUR POWER COMPANY.

3. Look for plumbing leaks and punctures in the roof that are wet or stained.

BASEMENT AND/OR CRAWL SPACE

1. Scan for loose wire and dangerous junctions.
2. Survey for condensation on HVAC pipes and any standing water at the system that is breeding mold.
3. Study the basement's earth, floors, and walls for wet spots and standing moisture that may indicate plumbing leaks.

Developing the Electrical, Plumbing, and Air Conditioning Checklist

The important step for getting a hold on maintaining these systems is to get some overview of what you want to achieve. The primary concern with all these systems is safety—there is not much chance to destroy other parts of the building with most of these systems.

Naturally the exception is water leaks from plumbing—they can be every bit as devastating as water infiltration from outside sources.

The Electrical, Plumbing, and Air Conditioning Checklist

Done	Item	Call a pro	Notes
✓	EXTERIOR	✓	
	TYPE OF SERVICE		
	ELECTRIC (Use a pro.)		
	Overhead from power line		
	Underground		
	Photovoltaic		
	Other		
	WATER (Use a pro.)		
	Underground at street		
	Private well		
	Type of water heater		

Done	Item	Call a pro	Notes
	Septic system		
	Other		
	GAS (Use a pro.)		
	Underground to meter		
	Propane		
	Other		
	HVAC (Use a pro.)		
	Type of heating equipment		
	Type of cooling equipment		
	Type of filtration equipment		
	Other		
	OTHER SYSTEMS		
	CONDITION OF SERVICE		
	ELECTRIC (Use a pro.)		
	I know how to turn off the electric service		
	I know how to operate my fuses or circuit breakers		
	My breaker boxes work well		
	Service wires in good shape		
	Power supply and meter box in good shape		
	Breaker boxes in good shape		
	Grounding system inspected by electrician		

Done	Item	Call a pro	Notes
	Ground fault interrupts installed in all places where needed		
	Other		
	WATER (Use a pro.)		
	I know how to turn off water service		
	No apparent water leaks in yard		
	Water heaters in good condition		
	Water heaters flushed out		
	Water heater overflow draining outside building		
	Solar water heaters maintained		
	Other		
	GAS (Use a pro.)		
	I know how to turn off the gas service		
	Pipes from meter to house in good condition		
	Other exterior gas pipes in good condition		
	Gas appliances in good condition		
	Other		
	HVAC (Use a pro.)		
	Heating unit in good condition		
	Cooling equipment in good condition		
	Heat pump units in good condition		
	Exterior filtration equipment		
	Other		
	OTHER SYSTEMS		

Done	Item	Call a pro	Notes
	INTERIOR		
	TYPE OF SERVICE		
	ELECTRIC (Use a pro.)		
	I know how to turn off electric service		
	I know how to operate my fuses or circuit breakers		
	My breaker boxes work well		
	Plate covers in good shape		
	Visible wiring in good shape		
	Ground fault interrupts in good shape		
	Other		
	WATER (Use a pro.)		
	I know how to turn off the water at entry to house		
	I know how to turn off water at appliances		
	No apparent water leaks in house		
	Shower doors not leaking water		
	Tubs and spas well caulked		
	All tile, and holes through tile for plumbing, sealed, grouted, caulked so they are not leaking		
	Water heaters in good condition		
	Water heaters flushed out		
	Water heater overflow draining outside		
	Other		

Done	Item	Call a pro	Notes
	GAS (Use a pro.)		
	I know how to turn off gas service		
	Pipes from meter to house in good condition		
	Other exterior gas pipes in good condition		
	Gas appliances in good condition		
	Other		
	HVAC (Use a pro.)		
	Heating unit in good condition		
	Cooling equipment in good condition		
	Heat pump units in good condition		
	Exterior filtration equipment		
	Other		
	OTHER SYSTEMS (Use a pro.)		
	LOW-VOLTAGE SYSTEMS (Use a pro.)		
	WARNING DEVICES		Bring the electrician out to review all battery needs—be meticulous in following instructions. It may be a good idea to have your electrician wire your entire home for you.

Done	Item	Call a pro	Notes
	Smoke alarms		
	CO_2 alarms		
	Radon detectors		
	Methane detectors		
	Other toxin detectors		
	Other		
	SECURITY SYSTEMS (Use a pro.)		
	Safe from water damage		
	Regular inspections set		
	BASEMENTS AND CRAWL SPACES		
	TYPES OF SERVICE		
	ELECTRIC (Use a pro.)		
	I know how to turn off the underground electric		
	All electrical work is safe from water		
	Water near electric system is not a threat to life		
	Wiring to code		
	Other		
	WATER (Use a pro.)		
	Water lines not leaking		
	Waste lines not leaking		

Done	Item	Call a pro	Notes
	No apparent water leaks below house		
	Water heaters in good condition		
	Water heaters flushed out		
	Water heater overflow draining outside building		
	All sumps and drain systems working		
	Other		
	GAS (Use a pro.)		
	I know how to turn off gas service below the house		
	Underground gas pipes in good condition		
	Gas appliances in good condition		
	Other		
	HVAC (Use a pro.)		
	Heating unit in good condition		
	Cooling equipment in good condition		
	Heat pump units in good condition		
	Interior filtration equipment		
	Other		
	OTHER SYSTEMS		

Done	Item	Call a pro	Notes
	ATTIC		
	TYPES OF SERVICE		
	ELECTRIC (Use a pro.)		
	I know how to turn off the attic electric work		
	All electrical work is safe from water		
	All penetrations of firewall are safe		
	Wiring in attic is not a threat to life		
	Wiring to code		
	Other		
	WATER (Use a pro.)		
	Water lines not leaking		
	Waste lines not leaking		
	No apparent water leaks in attic		
	Water heaters in good condition		
	Water heaters flushed out		
	Water heater overflow draining outside building		
	All penetrations in firewall safe		
	Other		
	GAS (Use a pro.)		
	I know how to turn off gas service to the attic		
	Attic gas pipes in good condition		
	Gas appliances in good condition		

Done	Item	Call a pro	Notes
	All penetrations in firewall are safe		
	Other		
	HVAC (Use a pro.)		
	Heating units in good condition		
	Cooling equipment in good condition		
	Heat pump units in good condition		
	Interior filtration equipment		
	Other		
	OTHER SYSTEMS (Use a pro.)		
	Experts for developing list are chosen		

 *Call the Doctor* 

Aluminum wiring can be dangerous. If you have aluminum wiring and don't know that it's okay or don't know if you have it or not, get it checked out by a seasoned, licensed, professional electrician.

Bringing It All Back Home

Air Conditioning

Building professionals call heating and air conditioning the mechanical systems of a building. This appears to be a very intricate array that provides the heating, ventilating, and air conditioning for a structure. In reality, mechanical components in the average home are not extremely complex. They typically consist of heating and cooling devices that condition air to a comfortable level and transfer it through pipes and grilles to the various parts of the home.

In older homes this can be as simple as a wood stove, a fireplace, a wall heater, radiators, or other devices that radiate out into the room. And in pre-air conditioning buildings, air control can be as straightforward as opening and closing windows and moving air with fans and attic ventilators to control it.

With later homes, the majority of systems have both a heating unit and a cooling device, which draw air from outside the building and adjust its temperature. After the air is conditioned for the inhabitants, fans circulate it out into the structure. The system has parts that give out, and there are many things going on behind the scene: Motor bearings are wearing, there can be off-gassing of toxins, corrosion exists, and chimneys clog. Numerous problems can develop with HVAC systems.

It is seriously recommended that you have your entire HVAC system checked by the correct professionals yearly—air cooling in the spring before the hot season kicks in and heating in late summer. Having your systems tight can save 10 percent or more on your energy bills and can add many years to the life of the equipment.

When systems are maintained, there can be a good deal of savings from not having to pay for excess fuel, continual and/or very expensive repair calls. And most important of all, the potential for hazards can be reduced considerably.

 **Call the Doctor** 

NEVER STAND ON A WET OR CONCRETE FLOOR AND HANDLE ELECTRICITY. GET AWAY FROM THE SITUATION AND HAVE THE POWER COMPANY INVESTIGATE. IN AN EMERGENCY, CALL THE FIRE DEPARTMENT.

POINTERS

Put only fuses in fuse boxes—never make substitutions. Make sure the fuses are of the correct amperage rating. If in doubt, get help.

 **Call the Doctor** 

Power lines on the ground should never be touched. Nor should you touch any surface such as a tree that is in contact with the lines or oven objects that are in the vicinity. Also remember that they can arc and jump. Just get away from them and make certain that the power company is aware of the problem.

Electrical

Are your circuit breakers tripping? Are fuses blowing? Are you standing in water or on a concrete floor while you use an ungrounded appliance? Are switches and plugs sparking? Are they cracked? Are there loose wires in the basement or metal electric boxes without covers on them? Is there plenty of clearance around your electric service and switch boxes? Are sinks and tubs and showers near plugs that don't have GFIs?

When you are having trouble with fuses or breakers blowing, or you don't know the answer to any of the questions above, it is important to have an expert out for a look-see. With electricity there is real danger, and it is never worth taking a chance.

Of course we are concentrating on how construction can continue to damage parts of the building, which is not a big part of electric work. For example, circuit boxes for fuses or breakers can rust through or cave in if they are poorly housed, but in the majority of situations, paying attention to safety and the lifetime of appliances is the main concern.

The system that provides electricity for a building is complex. There are many things going on which cannot be seen: grounding systems, corrosion of metal boxes, unprotected connections, magnetic fields, and joints of aluminum wires, just to list a few. A number of items can be dangerous, and parts of the system that are susceptible to deterioration can cause fire and, in worst-case scenarios, electrocution.

Because danger to the inhabitants is inherent with the electrical system, it needs to be monitored on a regular basis.

Plumbing

WATER

The water coming into the neighborhood through big pipes can leak and cause an entire street to become a sinkhole. Water running from the street to your house can leak and cause your lot to cave in or wash down a hill or can weaken the foundations of your home. The water branching out to water heaters, sinks, toilets, bidets, showers, etc., can leak and damage any part of your home.

Maintenance can take care of a number of these problems; for example, running drain cleaner through the drains monthly and a good flush when your plumber inspects once a year can prevent costly drain work. But this does not guarantee that roots or decomposition of pipes will not affect your home. Plumbing failures occur, and getting things fixed promptly can help to avoid compounding the situation.

As with the roof, have your plumbing, heating, and air conditioning inspected, cleaned, and maintained at least once a year. Many companies from the various trades offer professional inspections. A good construction attorney can review their insurance policies and the warranties the companies provide, so you know what you are getting into. You can also have an attorney draw up a contract or warranty and search for someone who will follow through.

Homeowners also should consider replacing ancient water heaters, toilets, furnaces, boilers, and other appliances with new high-efficiency or water-saving models. Most of us are in the habit of squeezing out every last day of service from such equipment. Yet, replacing any unit that is more than 12 to 15 years old will probably pay for itself within a few years through reduced energy or water usage.

As you prepare your plumbing checkup list, remember that the following are among the most common problems. Drains clog, especially at toilets, sinks, and tubs. The cause is usually what goes down the drains on a regular basis—problems can be avoided by not letting hair and grease and large objects go down the drain. Talk to your plumber about setting up a regular treatment with a rugged drain agent.

Faucet and handle leaks should be addressed promptly. Replacing old fixtures generally puts an end to problems.

If the piping system leaks, it can create a huge water and repair bill. Keep a close eye out during the maintenance operations. If a water heater is old, it may be time to just get a new one that is very energy-efficient—you might save the cost of the new unit via lower energy use in a short time.

It is easy for the cost of water slipping aimlessly through toilets that are in poor repair to break the $100 mark in a year's time. Toilets need work from time to time, so just keep up with it. This will prevent the occurrence of rot, mold, and termites at toilet areas, one of the most common locations for damage. The close neighbor to the toilet is showers and tubs—they are also big culprits in damage to homes, so keep an eye on them.

Gas

The gas system is, of course, very dangerous. Have your supplier take a look any time you are concerned. Gas does not tend to cause damage to the property. Water and building movement can, however, damage the gas system. During catastrophic events such as earthquakes, you must watch the entire neighborhood and vicinity for fire and explosion threats like gas leaks. This will call for keen attention to the situation after the event. It is best to do this with a group of people in your area who have experienced and understand emergencies and can take on various tasks to relieve one another. Make sure that everyone can turn the gas off at their homes.

If there is an event and your home appears unscathed, it is still wise to get the gas company to check it out afterward—when there is plenty of time—to be doubly sure. If there is a serious threat, abandon the property and call the fire department.

In General

The plumbing system is quite complex, with many things going on that cannot be seen: buried water supplies, waste lines which move sewage away from your home, flowing liquids, and gases. Many areas of the plumbing grid can be dangerous to the inhabitants and can undermine the integrity of the structure itself. Parts of the plumbing complex are susceptible to deterioration. A professional should check out questionable components, and maintenance should be attended to regularly.

As with electrical work, because danger to the inhabitants is inherent to the gas part of the plumbing system, it needs to be monitored on a regular basis. Also note that owing to moving water it receives a lot of wear and tear.

The Electrical, Plumbing, and Air Conditioning Resource Directory

Air Conditioning

airconditioner.com. Nationwide directory of air conditioner dealers, manufacturers, distributors, and HVAC groups. Locate extended-warranty companies.
http://www.airconditioner.com/

ASHRAE. See the American Society of Heating, Refrigerating and Air Conditioning for in-depth information about the HVAC world.
http://www.ashrae.org/

Clean up Your Air. Complete packages includes free shipping and and installation video.
http://www.alpinehomeair.com

DoItYourself.com—air conditioning and home heating repair. Learn about solar heating and air duct cleaning. Access guides and forums and check out the estimator for repair costs.
http://doityourself.com/hvac/

Heating/cooling pros near you. Central air, furnace, ducting, and more! Install or repair.
http://www.qksrv.net/click-1299592-10283446?SID=fw

HowStuffWorks.com. How Air Conditioners Work describes the way that this common household appliance cools interior rooms.
http://www.howstuffworks.com/ac.htm

HVAC City. Information from the Air Conditioning Contractors of America includes homepages for contractors, manufacturers, and distributors.
http://www.hvac-city.com/

Portable air conditioners and air coolers. CompactAppliance.com features everyday low prices and a great selection of portable air conditioners and air coolers.
http://www.compactappliance.com/jump.jsp?itemType=GATEWAY

Electric

APPA. This is the service organization for the nation's more than 2000 local publicly owned electric utilities. Full of excellent information.
http://www.appanet.org/

ASHI home inspections—electrical systems and hazards. Homeowners, buyers, and inspectors will find information about a variety of household topics, including a section with articles about electronics.
http://www.inspect-ny.com/#electric

Ask the builder—beware of do-it-yourself electrical repairs. Try to learn what you are doing before embarking on home electrical repairs.
http:www.askbuild.com/cgi-bin/column?452 (MSN)

Electrical-online.com. Electrical basics: safety, grounding, electrical code, electricity and your home (coming soon!). Common electrical terms and definitions.
http://www.electrical-online.com/howtoarticles/HowToIntro.ht

MSN House & Home, Electrical Systems. MSN House & Home offers information on home electrical components such as circuit breakers and fuses. Learn different ways to shut off the power.
http://houseandhome.msn.com/improve/electrical.aspx

Plumbing

Find expert plumbers with Rescue Rooter. Rapid response by uniformed, well-trained and licensed plumbing technicians. Services provided by ARS/Rescue Rooter. Online scheduling available.
http://services.servicemaster.com/tracker/FindWhat_Plumbing_.

Plumbing maintenance agreements. This is an interesting idea and probably a sign of things to come. This group in Los Angeles offers plumbing maintenance contracts. Once a year they will come to your home and perform a complete inspection.
http://www.priorityoneservice.com/PlumbingMA.htm

Plumbing maintenance—Service Magic—Prescreened contractors!
ServiceMagic.com gives a free match to prescreened, customer-rated plumbing contractors! Contractors may be in your area and interested in your job!
http://www.servicemagic.com

CHAPTER

9

Landscape and Hardscape

Most owners take a casual look at their landscape from time to time, even the busiest people who must spend long hours in the office. We all know the pleasure of looking out the window on a rainy day, checking out our grounds as the water falls, feeling cozy—the kind of thing that having a home is all about.

With a cursory glance from the study on a winter day, the grounds look very calm and peaceful. The yard, the plants, and watering systems are very complex and interrelated. There are many ways in which the earth beneath the plants affects the site: soils expand, mineral salt builds up, overwatering causes sogginess, berms block drainage, hillsides cave in, and so forth.

These situations can breed insects, cause rot, build soil up against the building and hardscape, and cause numerous other problems. Landscaping is a living, growing ecocenter that needs to be watched continuously and well cared for in ways that go beyond simple horticulture. The checklists cover maintenance needs that can help prevent repair work.

Hardscape refers to the built structures in the yard: retaining walls, gazebos, patio covers, sculpture, fences, gates, pools, spas, fountains, etc. Although it is not typical in professional construction classifications, we are including retaining walls (they are most often included in sitework) for the convenience of most homeowners who are not well versed in the building process.

Post in mud is a sure invitation for major repairs. Note that the earth and bricks create a basin that holds water against the post all year round—rain in wet season, sprinklers when it is dry. The bolts at the base of the post go through it and a metal strap. The strap is anchored in concrete below the surface to support and steady the frame. This is actually a small foundation like that of your house which holds the weight of the structure above. The construction defect is that the concrete foundation for the post does not extend 6 inches above the earth. The contractor did not pour the post base high enough, to allow for landscaping buildup. *Check your hardscape closely for soil buildup, just as you do the siding on your home.*

Rot at post base, even on a concrete surface. In this case the slant of the earth and the concrete are creating a basin that holds water against the post and the metal anchor. The post and anchor should be 6 inches above the concrete and should drain well. Rusting through the anchor, rot, and termites will result and could require thousands of dollars' worth of repairs if left unattended for long periods.

The hardscape can rot from soil buildup and suffer many other problems, just like the buildings. The hardscape needs to be watched systematically and maintained well. The checklists guide the owner through what to look for and how to set up the maintenance plan.

Jump-Starting

Landscape and hardscape are usually among the easiest parts of the parcel for the building owner to understand. Except for earth failure, which we looked at in Chap. 2, "The Building Site," the problems don't tend to be as complex as they are for the buildings. Let's take a look.

Six ongoing areas have requirements for tending to the landscape and hardscape:

FIGURE 9.3

Key through rusty fence illustrates the serious damage that sprinklers do to the hardscape as well as the home. If you find sprinklers flooding structures or the earth, they must be adjusted and moved if necessary. Then repair the damages and develop the strategy for tending to them regularly.

1. *Overwatering.* Examine automated devices; avoid flooding foundations and keeping structures wet.
2. *Drainage.* Keep water flowing away from structures and from the site.
3. *Damage from plantings.* Tree roots, plants against wood, and buildup of leaves and other cellulose by-products can damage buildings and hardscape. (Also see Chaps. 2 and 3.)
4. *Soils stability.* This topic is addressed in Chap. 2, "The Building Site," but its importance can never be overemphasized. Keep a close eye out for earth movement and buildup.
5. *Conditions of structures.* Monitoring the hardscape components is exactly the same as taking care of the main structures, but the scale is smaller.
6. *Pools, spas, fountains.* Three main things are important: Are they stable in the earth, are they leaking, and are they working well? Answering these questions requires a pro.

FIGURE 9.4

Metal fence rust can be very serious. Many metal parts of your hardscape as well as the major structures are highly vulnerable to complete disintegration in amazingly short periods of time. If your site water is charged with heavy mineral salts, it is just like salt on the road in cold areas. Very sturdy, baked-on, factory enamels on cars, as well as the protective coatings and the chassis itself, can be destroyed. The same is the case with your real estate. Even the nail heads on structural plywood inside your walls can rust and dissolve, leaving your entire building vulnerable to seismic and wind loads. Sprinkler overspray and poor caulk and paint practices cost homeowners millions of dollars each year.

The landscape and hardscape plan below gives some direction to gleaning the information for your own Landscape and Hardscape Checklist. With the plan in mind, we will move on to the "Quick Scan" section, walk your lot, and look at things outside before returning to the checklist. Remember not to get cavalier just because this chapter seems a bit easier than the others: earth movement, tree roots, and errant sprinklers—all the components here are important. If you don't know this material, seek professional coaching.

The Plan for Landscape and Hardscape Maintenance

The following outline applies to the problem of overwatering.

1. Review the "Quick Scan" section.

2. Consider all your watering closely.
   a. Areas that stay wet all the time
   b. Sprinklers spraying buildings and hardscape
   c. Algae and mold growth on overwatered structures
   d. Rusted buildings and hardscape
   e. Spalling and mineral salts deposits on concrete
   f. Trickles or streams eroding soils
   g. Water ponding
   h. Water standing in basements and crawl spaces
   i. Damp walls and soil in basements and crawl spaces
   j. Other suspicious situations.
3. Retaining walls
   a. Areas that stay wet all the time
   b. Sprinklers drenching foundations, v-ditches, and concrete work, placing a surcharge on the retaining wall
   c. Algae and mold growth on overwatered walls
   d. Rusted rebar exposed from effects of overwatering
   e. Spalling and mineral salts deposits on walls
   f. Trickles or streams eroding soils from around wall
   g. Water ponding at base or top or sides
   h. Walls, fractured or moving
   i. V-ditches and other surplus drains failing
   j. Earth slides at sides of walls
   k. Other suspicious situations
4. Structures
   a. Foundations wet all the time

FIGURE 9.5

All wooden parts of the hardscape (any structure on the parcel) are extremely vulnerable to rust, rot, and termites. Water must drain away from them. Note the fence boards above are embedded in the earth and already have algae blooming on them (the dark area near the dirt). The gatepost is buried in the earth and probably has concrete below that, and it carries the weight of the gate which puts a heavy, active load on it. If the post is treated for decay and concrete is aboveground with a slope away from the base for drainage, the unit will still rot in time, but the process will give the gate a much longer life span.

POINTERS

If the idea of hardscape is confusing, just go out and look at everything that has been built on your site that is not a part of the main buildings. That will get you started, and you can discuss it with the experts when you put your checklist system together.

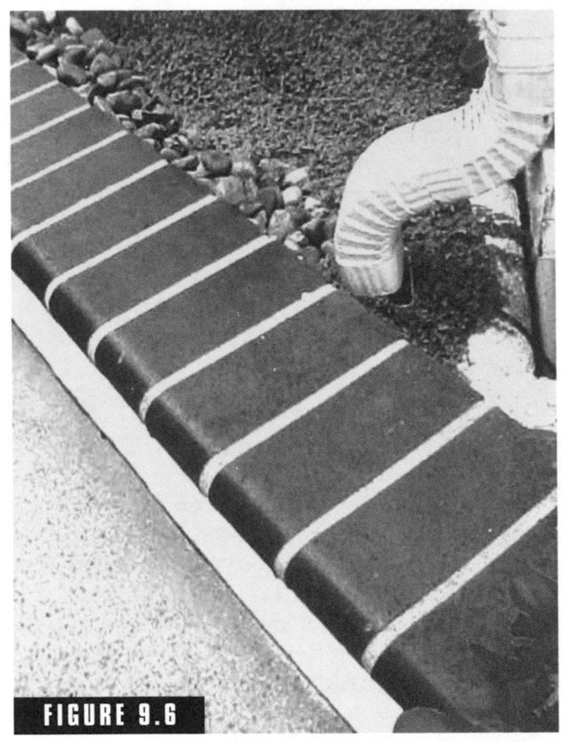

FIGURE 9.6

Downspout at wall next to side of house. Note that the spout itself is dislodged from the drain mouth that will take water off the site. The overflow of water can undermine the fence/retaining wall and even the foundation of the home, as well as cause sinkholes in the yard if the soils are not well compacted or there are vulnerable soils below the site grading that occurred during construction.

Call the Doctor

Any leaks, continually damp earth, and/or cracking of the basin or the concrete apron around pools or spas requires the immediate attention of a highly skilled professional. Never stall.

   b. Sprinklers drenching foundations
   c. Earth settlement
   d. Settling and structural problems
   e. Other suspicious situations
5. Pools, spas, fountains
   a. Foundations and aprons around them solid
   b. Water-holding parts of structures solid
   c. Leaks noticed in other areas of site
   d. Settling and structural problems
   e. Electric work, pumps, and filters working and safe
   f. Other suspicious situations
6. Miscellaneous
   a. Root disturbances—now and upcoming
   b. Plants affecting hardscape structures
   c. Earth disturbances
   d. Trickles or streams eroding soils
   e. Other suspicious situations
7. Start your checklist.
8. If you find any of the following conditions, get your expert immediately:
   a. Retaining wall problems
   b. Roots affecting critical areas such as foundations, gas, and electric
   c. Foundation problems
   d. Earth movement
   e. Watering system leaking and/or drenching buildings and hardscape structures
   f. Other suspicious situations
9. Discuss regular maintenance with the expert after any problems are fixed.

The Quick Scan

Below is the reference for monitoring the landscape and hardscape.

Quick Scan Outline—Landscape and Hardscape

As we stated above, you will be looking at six principal areas when you go out to examine the landscape and the hardscape:

FIGURE 9.7

Retaining wall damage can be severe even when the wall looks stable. Walls must be inspected every year after the rains, which can lead to some serious damage if a bank gives way suddenly.

FIGURE 9.8

Water trapped at fence and foundation caused by lawn flowing down toward the fence with the foundation right behind it. Pay close attention to areas where the landscape can affect both the hardscape and the main structures.

FIGURE 9.9

Leaky sprinkler stains planter. Pay close attention to anyplace in the landscaping or on any hardscape structures that has stains or stays wet. Leaks can cause damage that is very expensive to repair.

FIGURE 9.10

Site drains to patio—the landscape was not well coordinated with the layout of flatwork. Note that the sprinkler head in the corner next to the brick can easily keep the earth beneath the patio flooded through the dry season, and then drainage will continue to keep it flooded year-round. In time, this situation can allow the concrete to settle and cause the deep-down soil conditions below the subdivision's cut-and-fill work to be affected.

1. *Overwatering.* Examine automated devices and avoid flooding foundations and keeping structures wet.
2. *Drainage.* Keep water flowing away from structures and from the site.
3. *Damage from plantings.* Tree roots, plants against wood, buildup of leaves and other cellulose by-products can damage buildings and hardscape. (Also see Chaps. 2 and 3.)
4. *Soils stability.* This is addressed in Chap. 2, "The Building Site," but its importance can never be overemphasized. With pools and spas and fountains, large amounts of water surcharge the earth from leaks. Immediate attention from highly qualified professionals is required. Keep a close watch.
5. *Conditions of structures.* Monitoring the hardscape components is exactly the same as taking care of the main structures; even pools and spas are simple, reinforced steel structures. The big difference is that the scale is smaller.
6. *Pools, spas, fountains.* Check for leaks and stability.

First let's get in close to the house. You have walked it a number of times now, and some of it is probably getting familiar.

Although in most cases it would be much better if it weren't the norm, landscaping invariably surrounds the perimeter of buildings and hardscapes, growing up close to the walls. You walked it to look at the foundations and siding, and the concern was that plants should not be overwhelming the building—piling up at the foundation so moisture is trapped against the siding and growing too close to the building, flooding the siding and frame with water.

As you walk the beds again, notice how the plants relate to the structure. Are there large trees with massive trunks that will undermine the foundation in time? Are there rugged bushes that are growing right up against the siding? Is vegetation hugging the foundation and lower part of the siding, so the leaves will hold water against it?

Now, take a look at how the watering system for all the plantings around the building works. Are the sprinkler heads up against the foundation? Are they out in the bed and spraying the siding? Are mold and fungus growing on the siding at the foundation level? Are there signs of dirt that stays moist all the time?

While walking the flower beds, scan the lot as it surrounds the building. Does it slope down toward the structures? Is it flat? Does it slope down, away from the buildings? Do parts of it slope this way and other parts slope in other directions?

After you finish walking the plantings immediately adjacent to the building, tour all the rest of the lot, studying the basic questions: Is the watering system affecting the parcel negatively? Are the various hardscape structures in good repair? Are retaining walls working and stable? Are pools and spas tight?

When you sense that you have a bit of understanding of what to pull together for this checklist, it is time to do it.

FIGURE 9.11

Masonry walls can be easy to maintain. Although masonry can cost more than some fencing, as a product for walls and fences it is often one of the best buys in the marketplace. When factored over many years, well-engineered and well-placed block requires very little upkeep and repair.

Call the Doctor

If you see damp areas that appear to be related to the watering systems, do not wait—get an expert immediately. Surcharged water can cause serious damage.

POINTERS

Retaining walls are very important. Any major earth retention requires the use of licensed engineers both to analyze the problem and to design the retaining system. Any large retaining wall project requires the use of a contractor who is highly skilled in engineered earthwork.

Developing the Landscape and Hardscape Checklist

With a bit of understanding of how the landscape and hardscape relate to the entire parcel, let's start pulling together the checklist.

Landscape and Hardscape Checklist

Done	Item	Call a pro	Notes
✓	LANDSCAPE	✓	
	FLOWER BEDS		
	Plants against building		
	Roots disrupting foundation		
	Planting beds built up to foundations		
	Sprinklers soaking building		
	Damage from sprinklers on building		
	Other		
	YARD		
	Signs of sprinkler leaks		
	Wet areas		
	Other		
	HARDSCAPE		
	POOLS, SPAS, FOUNTAINS		
	Concrete apron around unit is solid		
	Cracks inside units		

Done	Item	Call a pro	Notes
	Signs of leaks around units		
	Earth moving around units		
	Other		
	RETAINING WALLS		
	Footings solid		
	Main body of wall solid, no cracking or shifting		
	Wall not leaking earth or water		
	Drains in wall working		
	Earth not leaking from sides		
	Other		
	OTHER STRUCTURES		
	Foundations solid		
	Buildup at framing		
	Sprinklers causing damage		
	Other		

Bringing It All Back Home

Care of the landscape is extremely critical to the long life of your home or any building you may own. Rot, mold, and insect infestations; foundation movement; and cracked concrete are among the most expensive repairs that building owners face. And remember that when the earthwork beneath a building fails and causes damages, the most expensive of all repairs are required.

Nothing can guarantee the life of your home, but diligent care can add to its longevity. A simple, ongoing landscape and hardscape maintenance program can save you hundreds of thousands of dollars in future repair bills.

Be sure to make the survey a habit. Do it three times a year—before, during, and after the rainy season. When you get to know the lot's runoff pattern, you will recognize any major problems that have come about from the storm season. Before the wet months, you may notice areas that have remained wet and piles of debris that can stop the flow. In the wet times, you can actually see slow and standing water. You can spot problems that should be fixed during the dry months. If you want to hand maintenance over to others, be sure to study the landscape and hardscape with them and listen closely to what they see. Not only does this put you in charge of maintaining your home, but you may have to change professionals in the future.

However, there is no guarantee that your efforts now will preclude intense work in years to come. Events like future development in your neighborhood can easily undermine the current drain system. The following are general thoughts for various types of lots—what to watch for and when to call the doctor.

As you take care of the tasks on your checklists through the years, watch for water that stands anywhere on the site, especially wet spots in dry seasons. Keep an eye out for any type of settlement like sinkholes anywhere on your site. Wet spots and settlement are immediate signals for calling a person you really trust. There are any number of things that could be causing these symptoms: broken water lines, broken drain systems, underground water. They are difficult to understand and often require boring, digging, and other exploratory work to find the cause.

Some people own sites that are adjacent to natural and man-made bodies of water. These parcels have their own unique characteristics and require the expertise of engineers and contractors who are familiar with the locale and the situation at hand.

Ocean and riverfront properties can be dramatically affected by the movement of water as watersheds drain land above into the large body of water. Man-made bodies of water can be defective, the liner of the bottom of the reservoir can leak, and many other problems can arise.

With all landscaping, the main thing to watch for is moisture that doesn't go away and earth movement—concrete lifting or breaking can be a strong indicator of earth movement. Any of these complex situations is reason to call in the an expert immediately.

Whether your environment encompasses hills, slopes, ditches, creeks, arroyos, rivers, lakes, or oceans, walking the neighborhood near your home will give you great insight. Just like in Chap. 2, do it in the rain, with your family. Note where the water goes: down the streets, into curbside storm drains, down the slopes, through your yard and the surrounding lots. The entire neighborhood could be undermining your foundation or causing mold and fungus to make your family have all sorts of allergies, discomfort, and even serious illness.

One of the most important aspects of developing a maintenance system is that you will become more and more familiar with your home and its environment. And just as in getting familiar with the sounds your car makes—any strange noise and you go to the mechanic.

The Landscape and Hardscape Resource Directory

ACRT, Inc. an employee-owned corporation. Tree condition, turf species, and hardscape maintenance are just a few of the categories of data you can collect. Green Resource Manager.
http://www.acrtinc.com/greenmgr.html

Compare estimates from local landscape contractors. Submit one request-for-quote form and get emailed rates and availability from up to five local landscape contractors. Free, no obligation service.
http://www.leadingcontractors.com/buyers/requestform.jsp?

Deluxe how to landscape. Gold Medal Designer's 2000-page book instantly online or CD-ROM.
http://www.internetgardens-us.com

Firewise landscaping: maintenance. Plants, planters, and landscaping maintenance.
http://www.firewise.org/pubs/fwl/mtn_frames.html

Gardener's News—gardening information. Besides advice articles, read success stories, ask gardening questions, and view archives. Topics rotate to include seasonal tips.
http://www.vg.com/gardening/

Hardscape, Akron, Ohio. Hardscape. Design and construction of patios, walks, waterscapes, and walls with unique materials. Hardscape is built for quality and longevity.
http://www.yourbackyard.us/hardscape.html

Landscaping. Home landscaping—landscaping burglars hate.
http://ci.lexington.ma.us/Police/CrimePrevention/landscaping

Landscaping for energy conservation. This site explains how trees and vines can be used to produce shade and protection from wind. It includes a list of species suitable for wind breaks.
http://aggie-horticulture.tamu.edu/extension/homelandscape/

Landscaping tips. Yard and garden advice from the experts at This Old House.
http://www.thisoldhouse.com/toh/yarda

Low-maintenance landscaping—explore MU Extension. The term *low-maintenance landscaping* should be kept in perspective.
http://muextension.missouri.edu/xplor/agguides/hort/g06902.h

Organic Land Care: program description. Organic Land Care practices.
http://www.organic-land-care.com/description.php

Planning the Home Landscape. Study a useful guide to landscape design and construction. It explains options, diagrams, materials, plants, and accessories.
http://aggie-horticulture.tamu.edu/extension/homelandscape/

Retaining walls. Restoration *maintenance.* Hardscape Renovations, Inc., is committed to providing professional and environmentally compliant hardscape restoration service through innovative equipment, skilled technicians, proven techniques, and quality materials.
http://www.trademarkconcrete.com/restoration.htm

ServiceMagic.com—prescreened home contractors. Get prescreened, customer-rated home contractors matched to your needs and interested in your job. Free service, no obligation, service guarantee.
http://ServiceMagic.com

Summer landscaping maintenance. Watering the lawn.
http://landscaping.about.com/b/a/005486.htm

Wonderfalls, Inc. Services include custom water features, hardscape construction, and landscape construction. Site offers pond and site maintenance, hardscapes, and sales of pond kits, products, and supplies. Photo gallery is featured.
http://www.wonderfalls.com/

Xeriscape: landscaping maintenance. Maintenance is one of the main seven principles of Xeriscape! In fact, regular maintenance is the key to success.
http://www.csu.org/files/general/2652.pdf

CHAPTER 10

Developing Your Personal Maintenance Plan

All the previous material in *Fix It Before It Breaks* has been aimed specifically toward this chapter. Now, we take your new knowledge about your home and distill it. As we stated from the beginning, the goal is a set of checklists. You will use them as an ongoing maintenance plan.

As we have said, the goal here is not to teach you to be a fix-it person, but to help you protect a large investment. See Fig. 10.1. As we have said, the process is very similar to providing service for your car—you don't have to learn to be the mechanic.

In that same vein of thinking, this book does not speak to ordinary subjects that are simply replacements of things that wear out or get used up. For example, take appliances. They just wear out; they have to be replaced, and they don't really affect other parts of the structure. There are a few exceptions such as toilets leaking and attracting mold infestations, but for the most part we skip completely over simple subjects such as door knobs, lightbulbs, or even paint that is simply cosmetic. It is similar to putting a blackener on your tires—you don't take the car to an expensive mechanic, and overlooking the work will not lead to more expensive repairs.

In most cases, your home will not be damaged if you never paint the interior walls. It may not look good to you or to others, but it will not really harm the structure. The subjects related to built structures are immense in number, and the vast majority of them, such as wall paint, are already covered in numerous books. Since there is no short-

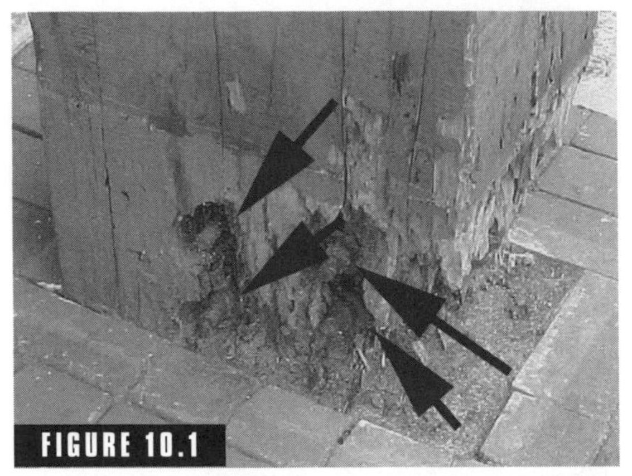

FIGURE 10.1

We showed you this photo at the beginning of Chap. 1, in Fig. 1.6. Now that you can see how important care is, we are showing it to you again. If the building had been built so the post sat well above grade on a concrete pad, and the trim boards had enjoyed diligent sealing and caulking, damage could have been avoided entirely. As you have learned, if the fault is not discovered and repaired as part of maintenance, then attended to regularly, hundreds of thousands of dollars of damage can be incurred—and you should remember that even bodily injury can result from structural failure.

age of information, we have narrowed the material to help the reader focus. We have not included any subjects that are not absolutely critical to the long life of your buildings.

You have covered quite a bit of information in this book, and it may seem a bit overwhelming to pull it all together. But you can just relax about getting things right and assume that it will all fall into place. However, remember that you can expect it to fall into place only if you put out the two safety lines that make that assumption work:

- You must use experts whenever you have any doubt or questions about damage that is taking place or getting worse around your property.
- You have to tend to the items on your checklists year after year (continuing to bring in professionals whenever you have doubts), which will automatically familiarize you more and more with the components of your real estate.

Never forget the old saw "The only dumb question is the one you don't ask." With the grave consequences resulting from the attrition that your home is always facing, the saying is especially true.

For the reader's convenience, to avoid repetition, and to get in rhythm with the types of wear and tear on your home, we have broken the majority of the ongoing plan down into seasonal maintenance tasks.

Review

Because we have provided the checklists for each of the sections in earlier chapters, you have some sense of how your home works as a group of systems. Let's go back through the previous checklists for the review. The goal is to look at how we can bring them all together for the final lists that you will continue to update and use through the years as you take care of your buildings.

Checklists

All of the lists guided you around your home, looking at how to achieve the maintenance goals for that chapter.

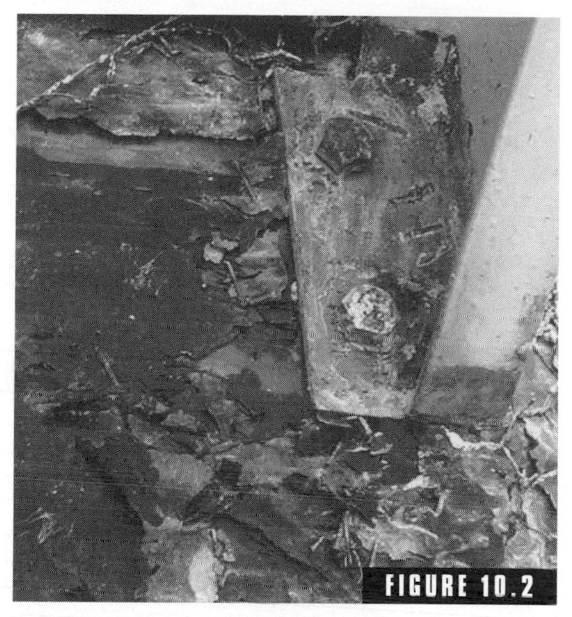

FIGURE 10.2

Even if you live in a condo, you are at the mercy of the maintenance team. If they are not well trained and diligent, the entire cluster of housing can be subject to millions of dollars in resultant damage. We also showed you this photo in the Chap. 1 as Fig. 1.3, and now you may have a better idea of how care can save you, and your Home Owners Association, money.

1 Getting Started

We do not integrate the "Who Does What Checklist" into seasonal checklists—they would become too long and clumsy. Keep it for all your needs: emergencies, adding on to your home, remodeling, as well as maintenance.

2 The Building Site

The two main site issues are

- *Drainage.* Make sure that surplus water flows away from your land readily.
- *Soil stability.* Monitor and control the movement, buildup, and/or erosion of earth.

3 Foundations and Cement Work

Remember where the dangers are with foundations and what you looked for related to your cement work:

- *Drainage.* Make sure that surplus water flows away from your land readily. Watch for damp spots and standing water.
- *Soil stability.* Monitor and control the movement, buildup, and/or erosion of earth. Watch for settlement and cracking.
- *Tree roots.* Survey for roots that are already disturbing your foundations and flatwork or that may do so in the future.

4 Framing

Watch out for

- *Shedding water.* The roof and siding must be tight. All joints and penetrations must be solid and well caulked, and paint must be in tip-top condition so water doesn't get to the frame.

FIGURE 10.3

Water leak at window jamb. Some of your best prevention work can take place as you clean and work around your home—stains on wall surfaces are one of the first indicators of problems.

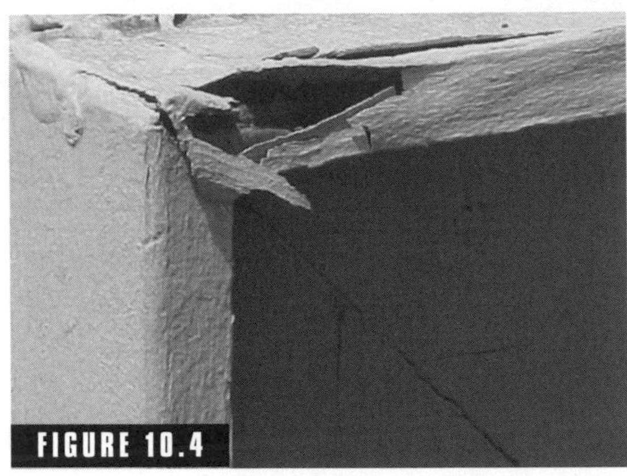

FIGURE 10.4

Just as for the interior of your home, exterior problems can be noticed while you are doing simple tasks such as raking leaves or watering. As you begin to know your property well, spotting problems can become second nature.

- *Earth movement and buildup.* Keep all soils and plantings carefully tended so they do not hold water against the bottom part of the framing.

5 Doors and Windows

These items need to be watched:

- *Protect the units.* Door and window units need to be cleaned, painted, and sealed on a regular basis.
- *Repair and lubricate working parts.* When doors and windows bind or are hard to work, don't shine it on, and do not slam them and beat on them. Lubricate them or fix them to keep the problems from getting worse.
- *Stop all water entry immediately.* Doors and windows are secured into the frame of the building with door and window jambs. The jambs are secured into a rough opening in the frame of the building. Both the surround of the door or window in its jamb *and* the crack between the door or window jamb and the siding must be well sealed and/or caulked at all times.

6 Siding

Take care of your siding:

- *Shedding water.* Siding must work as a solid membrane that continually sheds water.

- *All openings sealed.* This includes any holes, penetrations for vents, and the joints around doors and windows.
- *Paint or other membranes always maintained.* No matter what the final surface is (the first exterior skin)—paint, stains, coatings, stucco, etc.—they must not be cracking.

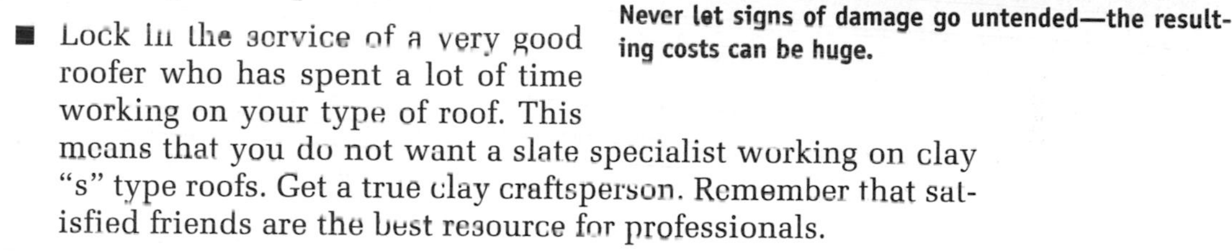

FIGURE 10.5

Never let signs of damage go untended—the resulting costs can be huge.

7 Roofs

Roofs need regular inspections:

- Lock in the service of a very good roofer who has spent a lot of time working on your type of roof. This means that you do not want a slate specialist working on clay "s" type roofs. Get a true clay craftsperson. Remember that satisfied friends are the best resource for professionals.
- Keep the roof inspected and clean, including the gutters, and use high-quality metal paint, such as Rustoleum, on galvanized flashing.
- Repair your roof promptly.
- Replace roofs regularly.

8 Electrical, Plumbing, and Air Conditioning

Keep an eye on them:

- *Water damage.* Make sure that water is not causing damage or promoting organisms at any part of these important systems.
- *System damage.* Check out your systems for any type of damage.
- *Maintain all the parts that require it.* Some parts of the systems, such as water heaters, require ongoing attention. Be sure you take care of them all.
- *Safety survey.* Check out all the systems for safety and maintain those components. If there are questions, get an expert.

9 Landscape and Hardscape

These are simple, but important:

- *Overwatering.* Examine automated devices and avoid flooding earth and foundations and keeping structures wet.
- *Drainage.* Keep water flowing away from structures and away from the site.
- *Damage from the plantings.* Tree roots, plants against wood, and buildup of leaves and other cellulose by-products can damage buildings and hardscape. (Also see Chaps. 2 and 3.)
- *Soil stability.* This is addressed in Chap. 2, "The Building Site," but its importance can never be overemphasized. Keep a close eye out for earth movement and buildup.
- *Conditions of structures.* Monitoring the hardscape structures is exactly the same as taking care of the main buildings, but the scale is smaller.
- *Pools, spas, fountains.* Three main things are important: Are they stable in the earth, are they leaking, and are they working well? (These questions are not a part of this book, but can be included as other maintenance.)

10 Developing Your Personal Maintenance Plan

Now you are in a position from which you can really begin to develop an upkeep system of your own.

Jump-Starting

If you are a person who prides yourself on being a quick study, you have probably picked up a lot of this system. We have attempted to make it as simple as possible. There are a great many books in the marketplace about every part of construction—anything related to building. There are plenty of books about the handyman parts of maintenance. But, as you know, this book is designed for the sole purpose of creating your personal checklists.

We are fully aware that we are getting monotonous with our repetition of this—but even if you are a dynamite jump-starter, call in the experts from your list if you have any questions. Don't even think about it, *just do it*.

That boils it all down for the jump-starter. Just review the work you have done, fill in the following lists, and you are off.

The monthly checklists are pretty much self-explanatory.

Home Maintenance Program—Keep an Eye Out Checklist

Done	Item	Call a pro	Notes
	THE BUILDING SITE		
	Walk your site regularly—look at everything: siding, roof, trees, flatwork. Watch for anything out of the ordinary. Do it with neighbors, and if you are using professionals, make certain that they are diligent.		
	FOUNDATIONS AND CEMENT WORK		
	I noticed:		
	FRAMING		
	I noticed:		
	DOORS AND WINDOWS		
	I noticed:		
	SIDING		
	I noticed:		
	ROOFS		
	I noticed:		
	ELECTRICAL		
	I noticed:		
	PLUMBING		
	Keep an eye out for leaky faucets—fix them fast.		
	Clogged drains—open them immediately.		
	MECHANICAL		
	Operable damper in the fireplaces.		
	LANDSCAPE AND HARDSCAPE		
	Wet spots		
	Earth movement		
	I noticed:		

Home Maintenance Program—Monthly Checklist

Done	Item	Call a pro	Notes
	THE BUILDING SITE		
	I noticed:		
	FOUNDATIONS AND CEMENT WORK		
	I noticed:		
	FRAMING		
	I noticed:		
	DOORS AND WINDOWS		
	I noticed:		
	SIDING		
	I noticed:		
	ROOFS		
	I noticed:		
	ELECTRICAL (May want to use a pro)		
	Check all the smoke detectors.		
	Check the sensor points for security system.		
	Check alarm for security system.		
	PLUMBING (May want to use a pro)		
	Check pool filters.		
	Check plumbing system filters: water softeners, purification, etc.		
	Run drain cleaner through all drains.		
	MECHANICAL (May want to use a pro)		
	Check heating and air conditioning filters, depending on season.		
	Clean the grill and condenser on AC unit.		
	Perform monthly work on heating and cooling systems.		

Done	Item	Call a pro	Notes
	LANDSCAPE AND HARDSCAPE		
	MISCELLANEOUS		
	Check all filters. (May want to use a pro)		

Home Maintenance Program Spring Checklist

Done	Item	Call a pro	Notes
	THE BUILDING SITE		
	Check all items from your list.		
	FOUNDATTONS AND CEMENT WORK		
	Check the basements and crawl spaces for areas that stay wet, have mineral salt deposits, rot, or termites—any signs of water damage, ponding, or running water.		
	Check all flatwork for disturbances, cracking, or other defects. Repair during summer including maintenance provisions.		
	Pressure-clean and seal if desired (can add to resale value).		
	FRAMING		
	Check for squeaky floors and bouncy stairs.		
	Check, repair, and apply sealants to all exterior wooden structures.		
	Check all termite flashing at house to structure connections, including fences.		
	DOORS AND WINDOWS		

Done	Item	Call a pro	Notes
	Perform an annual exterior inspection of your home's windows, particularly those facing extreme weather.		
	Scaling strips, putty, caulk, and whatever holds the glass in your windows must be in good shape at all times. Use a pro at first.		
	Drafts and air leaks at doors, windows, and heating and cooling systems throughout the house can cost a bundle. Use a pro the first time.		
	Wipe condensation that forms on the window frames during the winter months.		
	Check the operation of all doors and windows. Any sticking requires a pro the first time. Have them show you how to maintain.		
	Sealants at windows are very important. If there is bare wood, get a pro.		
	All paint at doors and windows must remain in good shape, every year. Use a pro at first.		
	Check that window and door frames are solid and free of rot. Any doubt or problems requires a pro.		
	Check energy units with a pro.		
	Look for water stains on floors, ceilings, and walls inside and outside at all doors and windows.		
	Look at exterior caulking around the outer edges of the window frame. Trim off any old, loose caulking and seal any gaps with a good-quality caulk		
	Check that all hardware (locks, opening mechanisms, etc.) operates smoothly		

Done	Item	Call a pro	Notes
	Make sure any exposed hardware screws are tightened securely.		
	Clean any sand, dirt, or dust from door and window hinges, sills, and tracks.		
	Check doors for smooth operation.		
	Other		
	SIDING		
	Check paint—anything suspicious, call the painter.		
	Clean siding as often as required.		
	Inspect all caulk; scrape, clean, and replace where needed and touch up where needed.		
	Check for cracks and penetrations, and seal them.		
	ROOFS		
	Clean gutters and downspouts.		
	Spring roof inspection (must include the attics) allows time to schedule repairs through the summer. Pro required.		
	Check all decks and other exterior sealed tops for flashing, cracks, and penetrations.		
	ELECTRICAL		
	Annual appliance service and safety inspection.		
	GAS		
	Annual appliance sevice and safety inspection.		
	PLUMBING		
	Run cleaner through drains for spring cleaning. May want a pro.		

Done	Item	Call a pro	Notes
	Clean and check all water heaters thoroughly. Learn from a pro.		
	MECHANICAL		
	Get ventilating and cooling systems ready for summer. Use a pro.		
	Clean clothes dryer vents and lint.		
	LANDSCAPE AND HARDSCAPE		
	Inspect all watering systems completely, including aim and adjustment of nozzles and controllers.		
	Ironwork and steel for breaks in paint and rust.		
	Fences—deterioration, need for paint and sealants.		
	Exterior structures—deterioration, need for paint and sealants.		
	Inspect and test all pool, spa, fountain, etc., equipment. Use a pro.		
	Check all wooden structures; repair and seal as needed.		

Home Maintenance Program—End of Summer Checklist

Done	Item	Call a pro	Notes
	THE BUILDING SITE		
	FOUNDATIONS AND CEMENT WORK		

Done	Item	Call a pro	Notes
	Check under house for mold, rot, termites, wet spots, ponding, streams, any indications of excess moisture. Find the cause. Fix and apply maintenance.		
	Check concrete flatwork for earth settlement, undermining by water, cracking, and any other disturbances.		
	FRAMING		
	Inspect walls for inordinate amount of nail pop-ups, cracks, bad corners, sagging ceilings or any other signs of serious problems developing in the building.		
	DOORS AND WINDOWS		
	Check weather stripping with wet and wind tests, and repair all weather stripping.		
	SIDING		
	Check all paint, caulk, mortar, grouts, sealants. Scrape, clean, and replace if needed.		
	Examine all surfaces for anything that is loose, for example, shutters.		
	ROOFS (Use a pro.)		
	Wet-test gutter and check gutter system for drainage, paint, clogs, damage in time for repairs.		
	Prepare for ice dam prevention.		
	Check all flashings for paint.		
	Check chimneys for water intrusion, including replacement of spark arrestors and chimney caps as needed.		

Done	Item	Call a pro	Notes
	ELECTRICAL		
	GAS		
	PLUMBING		
	Clean and test water heaters.		
	Check all pipe insulation.		
	Check toilets completely. Learn from a pro.		
	Check all under-sink areas for leaks.		
	Check all tubs completely including all caulk and sealants.		
	MECHANICAL		
	Check duct insulation.		
	Clean and test all heating equipment.		
	Check dampers and clean fireplaces, check screens and doors, and get complete professional inspection and flue and chimney cleaning.		
	LANDSCAPE AND HARDSCAPE		

Home Maintenance Program—End of Autumn Checklist

Done	Item	Call a pro	Notes
	THE BUILDING SITE		
	FOUNDATIONS AND CEMENT WORK		

Done	Item	Call a pro	Notes
	FRAMING		
	Prewinter tile check—all is grouted, caulked, and sealed tight.		
	DOORS AND WINDOWS		
	Take a quick check of all caulk.		
	SIDING		
	Take a quick check of all caulk.		
	ROOFS		
	Clean the roof and gutters and treat any roofing products that require it.		
	Paint all flashings that require it.		
	Recheck all roof decks, trellis, and patio covers for sealants and flashings.		
	ELECTRICAL		
	Prewinter check of all monitors and detectors: smoke, CO_2, radon, etc.		
	Prewinter low-voltage check: security systems, photovoltaics, etc.		
	PLUMBING		
	Check all gas pipe and connections, interior and exterior, for safety		
	MECHANICAL		
	Clean lint from dryers and exhaust.		
	LANDSCAPE AND HARDSCAPE		

Home Maintenance Program Midwinter Checklist

Done	Item	Call a pro	Notes
	THE BUILDING SITE		
	Clean all accessible drains		
	FOUNDATIONS AND CEMENT WORK		
	FRAMING		
	DOORS AND WINDOWS		
	SIDING		
	ROOFS		
	Clean all accessible gutters and drains		
	ELECTRICAL		
	PLUMBING		
	MECHANICAL		
	LANDSCAPE AND HARDSCAPE		

Bringing It All Back Home

Once these lists are completed, the fun begins. Don't just keep it all to yourself. Pass *Fix It Before It Breaks* around the neighborhood. Bring your friends over to look at what you are doing. Go over to their homes and discuss the similarities and the differences. This will give you invaluable lessons.

Don't think of maintenance as a chore that has you trapped. Think of it as a profit center that can steer any property owner clear of throwing away thousands of dollars.

If you are an extremely busy person or the type who doesn't like physical tasks, don't worry about it. Get your architect or contractor over to the house, and let that person tell you about your home. Pool your resources and save money. Have the neighbors over, and let the architect fill in all of you on the neighborhood drainage, the soils, and the various aspects of everyone's homes. You can share in the expense.

Take it all the way. Fun is very important in life, and the more fun you put into the process, the easier it will be to take care of your home on a continual basis. Make it a party each season. Have the whole block out and exchange notes. Hire the kids in the neighborhood to clean gutters and apply caulk. It is simple work and not that hard to learn. You may even find yourself enjoying tasks that you would have never dreamed that you would like.

If you are simply too busy for any of this, you can have your architect over with a maintenance firm or a handyperson and take care of all the details at one time. The architect and the handyperson will probably see things and have excellent ideas that you would have never even noticed, if left to your own devices.

Bringing in a variety of construction experts during the formation of the plan and at various times later on will enhance the project and bring it up to a serious level of professionalism. Use them for all the various seasons and all the lists in the book. You will probably cut down on their visits as time goes by, as you learn more about your home. Others, such as your roofer, will return year after year.

No matter who does what, the most important thing is that you now have a handle on what needs to be done, what isn't being done, and how to make certain about what should happen if you have doubt. You now have the opportunity to take the reins for guiding the welfare of your buildings.

After knowledge of what needs to be watched on your home, the next most important thing is consistency. Never let it slide, and you will have most probably saved your estate a lot of hassle and money. Each year, walk the neighborhood with your neighbors and survey any changes in the drainage system that may have taken place: new homes, subdivisions, changes on hillsides, clogged drain areas, slides, sinkholes. Even if they are not on your site or immediately adjacent to it, all local conditions can impinge on your home.

Every 2 or 3 years, it would be beneficial to have your builder or your architect walk the site with you again for some overview. And don't forget to get together with your neighbors; you can pool the architect expense, and they may notice things that you are not seeing.

INDEX